The Equations of Life

How Music, Ageing, and Cancer Obey the Same Math

Dr. Sasi Shanmugam Senga

KCF Press
Boston • London
Newark, DE 19713, USA

KCF PRESS

The Equations of Life
How Music, Ageing, and Cancer Obey the Same Math

Published by KCF Press
Boston • London
KCF Press is a nonprofit publishing imprint.

This work was independently authored and editorially reviewed prior to publication.

First published 2026
ISBN 978-0-6489285-2-2

A catalogue record for this book is available from the Library of Congress.

The moral right of the author to be identified as the author of this work has been asserted in accordance with the Copyright, Designs and Patents Act 1988.

This book is intended for informational and conceptual purposes only and does not constitute medical advice. It should not be used as a substitute for professional medical judgment, diagnosis, or treatment.

Stability is not permanence.

Contents

Chapter 7
Phase Transitions in Living Systems
When Behaviour Changes Regime
Interlude 9: Diffusion beyond biology

Chapter 8
Health as Dynamic Balance
Stability Without Stillness
Interlude 10: What Mathematics Cannot Save

Chapter 9
The Equations of Life
What Persists, What Drifts, What Breaks
Interlude 11: Markets fail for the same reason bodies do

Epilogue
Beyond Life
Quantum Systems and Artificial Intelligence
Stephen Hawking: Structure Without the Body

On Living Against Fixed Odds

Appendix

Mathematics Under Constraint
When Time Determines Outcome

- The Atlantic, 1942
- The Manhattan Project
- Apollo 11
- Civilisation as a Regulated System

Notes

Author's Note: How to Read This Book

This book is not a textbook, and it is not a manifesto.
It does not propose a new theory of life, nor does it reduce
biology, medicine, or music to equations in the ordinary sense.

It is an attempt to make visible a set of mathematical constraints
that recur wherever systems persist under pressure.

Across disciplines that rarely speak to one another, the same
structures appear. Regulation stabilises behaviour. Noise
introduces variation. Time accumulates consequence. When
constraint holds, coherence endures. When it weakens, systems
drift. When it collapses, behaviour changes regime. These
patterns are not metaphors. They are properties of dynamical
systems realised in different material forms.

I have written this book as a neurosurgeon, and as a scientist
trained to recognise structure beneath complexity. My interest is
not in calculation, but in limits: what can be preserved, for how
long, and at what cost.

The equations that appear in these pages are not tools to be
used, solved, or applied. They are anchors. Each compresses an
idea that recurs across music, living systems, ageing, and disease.
They are included not to calculate outcomes, but to stabilise
meaning. No symbol is required to be remembered. If an
equation feels opaque, you may pass it by. Its role is to hold
structure in place, not to test fluency.

Mathematics does not replace meaning, experience, or care.
Meaning arises within constraint, experience is shaped by

regulation, and care is effective only where control can still be exercised. Nothing here promises permanence. Regulation is costly. Balance is temporary. Drift is unavoidable. Understanding does not confer immunity.

What it can confer is precision.

The examples in this book are drawn from music, physiology, ageing, cancer, history, and technology. They are chosen not for novelty, but for clarity. Each exposes the same logic under different conditions of noise, delay, and control. Historical episodes and contemporary systems appear not as illustrations, but as stress tests. They show what mathematics does when error is unforgiving and time matters.

The chapters follow a deliberate progression, but they do not demand strict linear reading. Music, life, ageing, and cancer are explored as different regions of the same mathematical landscape. Later chapters may clarify earlier ones. This is not a flaw in reading. It reflects the subject itself.

Interludes are not explanations. They are pauses. They address what it feels like to live inside regulated systems: to anticipate, to drift, and to approach limits. They may be read quickly or slowly, or returned to later.

This book makes no promise of rescue. It distinguishes between delay and control, between stability and appearance, and between what can be influenced and what cannot. Its aim is clarity, not consolation.

The purpose of this book is to make certain patterns easier to recognise: when systems are stable, when they are merely delayed, and when they are approaching limits beyond which recovery is no longer symmetric.

Nothing here depends on memorisation. What matters is recognition.

This is a book about what persists, what drifts, and what breaks.

Read it as you would listen to music: not to master it, but to hear when coherence holds, and when it does not.

Orientation: The Mathematical Landscape of This Book

This book is organised around a single mathematical idea: that coherence depends on structure, regulation, noise, and time.

The different domains explored here correspond not to different metaphors, but to different regions of the same mathematical landscape.

• Music illustrates strict constraint with immediate feedback.
• Life illustrates regulated coherence under unavoidable noise.
• Ageing illustrates drift as regulatory precision erodes.
• Cancer illustrates regime change when constraint collapses.
• Health illustrates bounded behaviour without stillness.
• Historical and artificial systems illustrate limiting and contrast cases.

The mathematics does not change across these domains. What changes is the balance between constraint, correction, noise, and time.

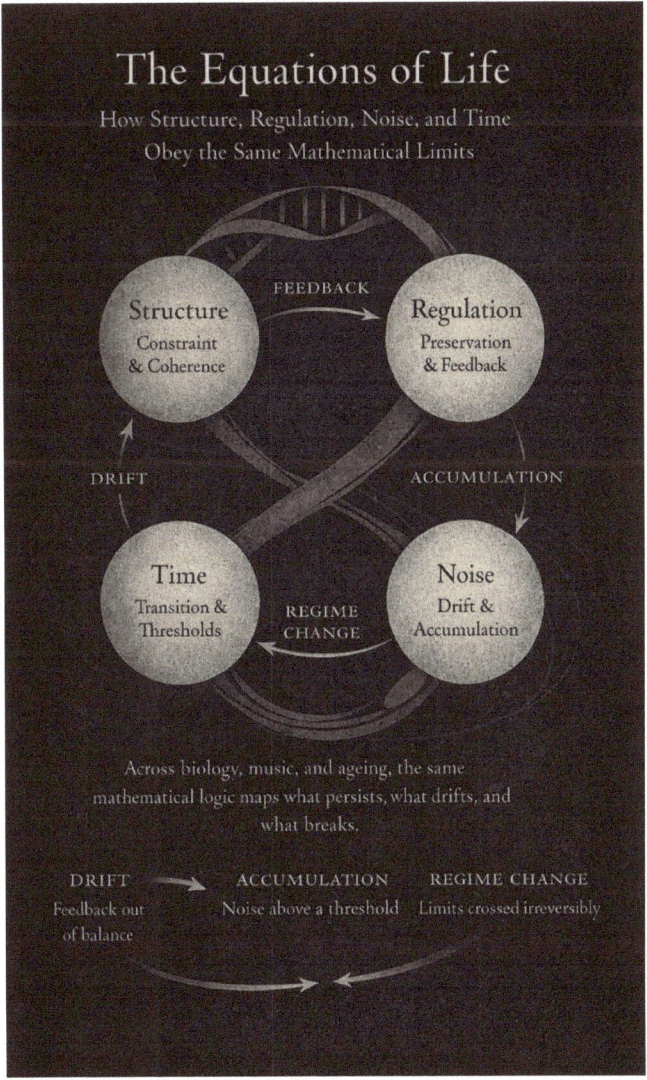

The Equations of Life

How Structure, Regulation, Noise, and Time Obey the Same Mathematical Limits

Structure
Constraint & Coherence

FEEDBACK

Regulation
Preservation & Feedback

DRIFT

ACCUMULATION

Time
Transition & Thresholds

REGIME CHANGE

Noise
Drift & Accumulation

Across biology, music, and ageing, the same mathematical logic maps what persists, what drifts, and what breaks.

DRIFT
Feedback out of balance

ACCUMULATION
Noise above a threshold

REGIME CHANGE
Limits crossed irreversibly

A conceptual map

About the Author

Dr. Sasi Shanmugam Senga is a neurosurgical oncologist with a Master's degree in Neuroscience and a Master's degree in Cancer and Therapeutics. He is a UK Commonwealth Scholar, an Oxford Clarendon Scholar, and a recipient of the Harvard Excellence Award.

He has served as a lecturer in Medicine at the University of Oxford and the University of Buckingham, and as Programme Director in Molecular Genetics and Ethics at Stanford University. He has also contributed to international policy and research discussions as a panellist for *The Economist* think tanks.

In parallel with his academic work, he is involved in the evaluation of global higher-education systems and contributes annually to the QS World University Rankings for top universities worldwide.

He is an Ambassador and active member of leading international cancer organisations, including the European Association for Cancer Research, the European Society for Medical Oncology, the American Association for Cancer Research, and the American Society of Clinical Oncology.

Dr. Senga is the author of a Royal Society–top-cited research article, *Hallmarks of Cancer: The New Testament.* His work lies at the intersection of cancer biology, ageing, neuroscience, ethics, and the limits of medical intervention.

In memory of Kalavathi

mother of the author,
whose life and death shaped the author's path toward
contributing, in however small a way, to the benefit of
humankind.

Before the mathematics, there is a life lived inside its limits.

A violin string vibrates and music emerges.
Human Prelude
Hospital corridor → text → surgery → temporary order

The hospital had a way of making time feel both urgent and repetitive. Doors swung open, gurneys passed, monitors chimed in the same narrow register of insistence. Every day contained its own emergencies, and yet most days looked the same.

I was still early enough in training to believe the sameness was a kind of promise. If I could learn the pattern well enough, I could intervene sooner. If I stayed longer, read more, asked the right questions, the outcome would change.

It wasn't a noble idea. It was a practical one. The whole institution was built around it.

I had been in theatre since before sunrise. The air always ran cold, calibrated for sterile comfort rather than human warmth. Under the lights, everything became abstracted into fields: skin, fascia, bone, the clean geometry of an incision. Outside the drapes, the patient was a person. Inside them, they were a problem to be solved.

Between cases, you came up for air.

I stepped into the corridor that ran along the operating suite. The doors were heavy and soundproofed; the moment they closed, the world changed. The corridor smelled of antiseptic and reheated coffee. The fluorescent lights made everyone look slightly ill.

I took off my gloves and peeled my cap back from my forehead. The indentations from the elastic sat in the skin like temporary scars. My phone was in my locker; we didn't carry it in theatre. In those years, I still checked it with the superstitious fear that if I missed something, I'd be responsible for whatever followed.

I opened the locker, scrolled with one thumb, half-thinking about the next patient on the list, what I'd read the night before, what I'd decided might be worth trying differently.

Then I saw my mother's name.

It was a text message. A few lines. Clinical, compressed, written the way people write when they do not yet have words large enough for what they are trying to contain.

Brain metastases.

There are phrases that do not behave like language. They behave like impact. You can understand them instantly and still need time to catch up to the fact that you understood them.

My training made the words expand automatically. I saw images I hadn't been shown yet. I populated the sentence with lesions, locations, probabilities. I heard the voice I used with patients and families: measured, careful, structured around what could be done.

And beneath that, a simpler recognition arrived, uninvited.

This is my mother.

Someone called my name down the corridor, a casual shout over the hum of ventilation. A nurse laughed at something in a doorway. A trolley rattled past. Nothing in the building registered that the axis of my life had shifted, because nothing in the building was built to register that.

I stood with the phone in my hand longer than I meant to, staring at a message that did not change no matter how many times I read it.

I could tell you exactly what I felt. But that would be dishonest in a different way, because what happened first was not a feeling. It was a sequence.

I walked back into the changing area. I washed my hands again though they were already clean. I re-tied my mask. I went back into theatre.

The case proceeded. Instruments passed into my palm. Tissue responded the way tissue responds. The body did not know what had arrived on a screen in a corridor.

When the case ended, the list still existed. The hospital still had its appetite. Patients still arrived in batches, each one holding the private conviction that their crisis was singular, and each one correct.

That evening, I did what I had been trained to do: I made plans.

Brain metastases is not a diagnosis so much as a declaration of context. It tells you that the disease is not contained. It tells you that even perfect local control is still local. It tells you that you can win battles and still lose the war.

But it also tells you something else, if you are willing to look closely: there is still a narrow space where intervention matters.

I found the right people. I arranged the flight. I moved my mother through the channels of expertise and urgency that medicine reserves for the cases that might still bend.

When the time came, I insisted on joining the surgical team, not as a gesture, and not because it would change the biology, but because the act of standing at the table is sometimes the only way to endure what you already understand.

The tumour came out. The brain, relieved of pressure, did what brains sometimes do when given the chance: it returned function.

She stood.

Then she walked.

It is hard to describe the feeling of watching someone you love walk again after you have already rehearsed their decline. The body believes what it sees. The mind knows what it knows. For a moment they live in different worlds.

She lived for several more years. Long enough for routines to re-form around her. Long enough for ordinary conversations to return. Long enough for the reprieve to feel like something other than borrowed time.

The story I had been telling myself since my intern years, that with enough attention and enough knowledge you could fix the problem, did not shatter all at once. It simply stopped being adequate. I would later recognise the same pattern in other lives.

This book begins there: with a temporary restoration of order, and the quiet knowledge that order is not the same as permanence.

A Case of Apparent Stability

What follows is a different life, governed by the same constraints. She was sitting alone when I opened the door, coat still on, bag at her feet. The room smelled faintly of disinfectant and coffee gone cold. She looked up immediately, half-smiling.

"I'm sorry," she said before I spoke. "I know you're busy."

She was in her early forties. Her phone lay face down on her lap. She glanced at it once, then pushed it slightly out of reach. Later she would tell me she had an alarm set, school pickup, a fixed point in the afternoon she could not miss.

Her symptoms had not announced themselves. They had accumulated. A headache that lingered when others faded. Words that occasionally arrived late. A tiredness that felt different from being busy, but was easier to explain that way.

The scan removed any ambiguity. The lesion was unmistakable, but contained. No spread. No urgency. The kind of finding that allows sentences like *we have time* to be spoken aloud.

She studied the image quietly. When she spoke, her questions were precise. How large. Where exactly. What next. She did not ask whether it would change her life. That question did not need to be voiced.

We intervened.

The operation was uneventful. Pressure was relieved. Signals settled into familiar patterns. In recovery, she opened her eyes and spoke clearly. She followed commands. There was a restrained sense in the room, shared, careful, that the worst had been averted.

In the weeks that followed, she resumed her routines cautiously, as if testing the ground before committing her weight. She learned which days required rest, which could stretch a little further. She became precise with time, with meals, with sleep.

Follow-up imaging was stable. No new findings. No obvious progression. The language in clinic remained optimistic. *Recovering well. Encouraging signs.*

But something else was happening.

She described it without drama. How a late night now echoed for days. How a missed meal left her unsteady. How concentration sometimes thinned without warning, then returned, leaving her unsure whether to trust it.

Nothing had failed.

The interventions still worked, but they worked differently now. Each correction brought her close to where she had been, but not quite back. The centre she was correcting toward had begun to move, slowly and without signal.

From the outside, she remained capable, articulate, independent. She still arrived on time. Still spoke clearly. Still met expectations. From within, everything required calculation. Stability was no longer ambient; it had become labour.

Over time, small deviations began to leave traces. Variability widened. The space for error narrowed. What once disappeared quietly now accumulated.

There was no moment when this could be named as decline. No threshold crossed. No scan that justified alarm. The notes remained careful. *Managed. Compensated. Under control.*

But control was no longer symmetric.

Eventually, often recognised only later, the familiar interventions stopped returning her to the same place. Not because they failed, but because they were now correcting against a moving reference. What could be restored was no longer the original state, but the closest version still sustainable.

This was not collapse.
It was not loss of regulation.

It was the cost of maintaining order over time.

She adapted. She reorganised. She learned where the edges lay and stayed clear of them. Life continued, recognisable, functional, but under tighter constraint.

Medicine gives names to this: progression, ageing, inevitability. But those names conceal what is actually happening.

The system persists, but at a higher cost.
Coherence remains, but with less room to move.
Time has simply been allowed to accumulate.

Introduction

Listening for the Mathematics beneath life

A violin string vibrates and music emerges.
A heart cell oscillates and life continues.

At first glance, these events appear unrelated. One belongs to art, the other to biology. Yet both persist for the same reason. They are constrained. They obey structure. What we recognise as harmony in one case and health in the other arises not from chance, but from regulated mathematical order.

This book will give you a way to recognise when a system, musical, biological, or human, is truly stable, when it is merely holding on, and when it has crossed a point beyond which recovery is no longer symmetric.

A heart rate that rises under exertion and returns without thought reveals constraint long before it reveals number. Mathematics is often mistaken for calculation. In its deeper sense, it is the study of form, of what can repeat without collapsing, of what remains stable under disturbance. It governs rhythm, balance, feedback, and failure. Wherever behaviour is coherent rather than arbitrary, mathematics is already at work.

This book begins from a simple observation. The same mathematical patterns recur across very different domains. Oscillations, feedback loops, thresholds, noise, and instability appear in music, in living cells, in the gradual changes associated with ageing, and in the abrupt loss of control seen in cancer.

Music offers the clearest case. A string fixed at both ends can vibrate only in particular ways. Those permitted vibrations produce harmony. Order emerges not because the system is expressive, but because it is constrained. Music is mathematics realised under conditions of near-perfect obedience.

Life is governed by the same mathematics, but under harsher conditions. Living systems operate far from equilibrium. They fluctuate, repair, anticipate, and adapt. They survive not by eliminating randomness, but by regulating it. Here, mathematical order does not disappear. It becomes probabilistic, distributed, and continually corrected.

Over time, that correction loses precision. Small errors accumulate. Rhythms drift. Signals grow noisy. The system remains functional, but less exact. Ageing is not a programme of destruction. It is the gradual erosion of regulatory accuracy.

Sometimes regulation fails more abruptly. Control mechanisms that once stabilised growth and repair begin to amplify deviation. Feedback turns self-reinforcing. Constraint gives way to instability. Cancer does not arrive from outside life. It emerges when the mathematics that normally sustains organisation is driven beyond its stable range.

Across these domains, the differences are profound. Yet the underlying logic is shared. Each can be understood as a dynamical system operating under different conditions.

Mathematics does not replace biology, emotion, or meaning. It is the framework within which they arise. Music allows us to hear stable form. Life allows us to observe regulation under uncertainty. Ageing reveals the cost of time. Cancer exposes what happens when control is lost.

Life is not written in spite of mathematics, but because of it.

To understand life more deeply, we must learn to listen. Not only to sound, but to pattern. To rhythm. To deviation. To the moments when structure holds, when it drifts, and when it finally breaks.

This book is an invitation to listen more carefully. If the patterns described here are real, then many of our intuitions about life and disease are incomplete.

We are accustomed to thinking of health as a state that can be restored, ageing as damage that can be repaired, and cancer as a malfunction that can be eliminated. These assumptions are not irrational. They follow naturally from a focus on parts, interventions, and short-term success.

But systems governed by mathematical structure do not fail in this way; they drift, destabilise, and cross thresholds. They allow local repair without global control, delay without reversal, and improvement without cure.

Ageing and cancer are often treated as distinct biological failures. In what follows, they will be examined as mathematically adjacent outcomes of the same regulatory architecture.

To understand life through this lens is to accept that some forms of breakdown are not errors to be corrected, but consequences of how order is maintained over time.

What follows is not a rejection of medicine or biology, but a reframing of what they can reasonably promise.

Chapter 1

Mathematics as Pattern

Before mathematics became a tool for calculation, it was a way of recognising form. Long before equations were written down, patterns were observed, compared, and trusted. Day followed night. The seasons returned. A vibrating string produced the same pitch each time it was plucked in the same way. Mathematics began not as abstraction, but as the disciplined recognition of regularity.

At its most fundamental level, mathematics is concerned with **structure**. By structure we mean the constraints that determine what a system can do and what it cannot do. Mathematics describes what may vary and what must remain invariant. It defines the boundaries within which behaviour is possible. In this sense, mathematics does not impose order on the world. It records the order that already exists.

Structure is therefore not appearance, but permission. A system with structure does not evolve arbitrarily. It moves within a defined space of possibilities. Some behaviours are allowed. Others are forbidden. This distinction is what makes coherence possible.

Throughout its history, mathematics has advanced not only through formal proof, but through a persistent sensitivity to structure that often precedes justification. This sensitivity has taken many forms. In some cases it appears as immediate insight, in others as prolonged obsession, and in others still as an ability to recognise order where none seems obvious.

Few figures illustrate this more clearly than **Srinivasa Ramanujan**, who, working largely in isolation and with little formal training, produced mathematical results of extraordinary depth. Ramanujan described many of his formulas as revelations, arriving fully formed rather than derived step by step. When he sent a letter of largely unproven theorems to Cambridge in 1913, they appeared so unfamiliar that they were initially suspected to be fraudulent. They were not. Many were later proven rigorously, while others continue to inform modern mathematics and physics. His work demonstrates that mathematical structure is often recognised before it is formally secured.

Ramanujan himself struggled to describe how this recognition occurred. He attributed many of his discoveries not to deliberate derivation, but to sudden appearance. He famously said that his equations were given to him in dreams, arriving complete rather than assembled.

In his own words, he described these moments as gifts from the goddess Namagiri of Namakkal, who, he said, placed mathematical formulae on his tongue while he slept. "An equation," he wrote, "has no meaning unless it expresses a thought of God."

It is tempting to read such statements either as mysticism or metaphor. Both miss the point.

What Ramanujan was describing was not the suspension of structure, but its *prior apprehension*. The mathematics was not unformed. It was simply not yet articulated in the language of proof. Cultural tradition provided him with a vocabulary for an experience that mathematics itself would later formalise.

His results were correct not because of their origin stories, but because they obeyed structure. Proof followed insight, not the other way around.

Ramanujan's case makes something explicit that is usually hidden: mathematical understanding often begins as recognition before it becomes justification.

A different expression of the same phenomenon appears in the work of **Carl Friedrich Gauss**, whose genius lay not in speed of calculation, but in the immediate perception of structure. As a schoolboy, Gauss famously summed the numbers from one to one hundred almost instantly by recognising a hidden symmetry. The episode is mathematically elementary, but conceptually decisive. Mathematical power lies not in labour, but in seeing constraint.

The primacy of structure over calculation became explicit in the work of **Evariste Galois**, who showed that the solvability of equations depends not on algebraic manipulation, but on symmetry. Writing in extraordinary haste on the night before his death at the age of twenty, Galois laid the foundations of group theory. His insight revealed that behaviour is governed by underlying structure rather than surface complexity.

In the twentieth century, this idea reached its most profound expression in the work of **Emmy Noether**, who demonstrated that the laws of physics are inseparable from mathematical symmetry. Conservation laws, long treated as empirical facts, were shown to arise from invariance. With Noether, structure ceased to be merely descriptive and became explanatory.

This is not an abstract insight about equations; it is the reason living systems can remain recognisable while everything inside them changes. Structure matters because it constrains behaviour. A system governed by mathematical structure cannot change arbitrarily. Whether the system is a musical instrument, a living cell, or a population of dividing cells, its behaviour is shaped by rules that determine stability, repetition, and permissible change.

The importance of recognising such structure is reflected in how mathematics itself is valued. The **Fields Medal**, often regarded as the highest honour in the discipline, is awarded not for utility, but for insight into deep and unifying structures. It recognises work that reveals unexpected connections across disparate domains, precisely the kind of understanding required to approach complex systems.

One of the simplest examples of structure is oscillation. A system displaced from equilibrium returns, overshoots, and returns again. This behaviour appears wherever opposing forces interact. In sound waves, oscillation produces pitch. In physiology, it produces heartbeat and respiration. In cellular biology, it governs cycles of growth and division. The material differs. The structure does not.

Another fundamental form of structure is feedback. Negative feedback stabilises systems by counteracting deviation. Positive feedback amplifies change. These are not metaphors, but formal properties of dynamical systems. When negative feedback dominates, systems remain robust. When positive feedback overwhelms control, instability follows. The same structural logic governs a musical performance, a metabolic pathway, and the regulation of cell growth.

Mathematics also clarifies the limits of structure. Some problems resist solution not because of technical difficulty, but because they expose fundamental boundaries. Questions concerning prime number distribution, the nature of randomness, or the behaviour of nonlinear systems remain open despite centuries of effort. Such problems are not failures of mathematics. They mark the edges of what structured understanding can currently reach.

Crucially, mathematics distinguishes between linear and nonlinear structure. Linear systems respond proportionally to change. Nonlinear systems do not. Small inputs may produce large effects, and large inputs may have little impact. Living systems are inherently nonlinear. They rely on thresholds, switches, and interactions across scales. This nonlinearity allows adaptation, but it also introduces fragility.

Mathematics shows that order and disorder are not opposites. They are related states within the same structured system. Excessive rigidity prevents adaptation. Excessive variability destroys coherence. Functional systems exist within a narrow region between these extremes. This region can be described, measured, and modelled.

For this reason, mathematics provides more than description. It provides expectation. When a system violates its expected behaviour, something has changed in its underlying structure. Mathematics tells us not only what happens, but what should happen, and therefore when regulation has failed.

Throughout this book, mathematics will appear not as a collection of equations, but as a unifying logic. The same structures recur across different domains. In music, they produce harmony. In life, they enable regulation and adaptation. In ageing, they drift. In cancer, they destabilise.

What differs is not the mathematics itself, but the conditions under which structure is maintained.

To understand life in its many forms, we must first understand the structures that make behaviour possible at all. Mathematics is the study of those structures. Everything that follows depends on this foundation.

Interpretation

Mathematics is often mistaken for calculation, yet its deeper role is the identification of constraint. It reveals what is possible, what is stable, and what cannot endure.

In life, coherence is never accidental. What persists does so because it is structured in the right way. Freedom without constraint dissolves. Constraint without flexibility petrifies.

To recognise structure is therefore to recognise the conditions under which life can hold together at all.

A further distinction follows from this view of structure.

Systems may exhibit improvement without preserving structure. Function can return locally while invariants continue to erode globally. Time may be gained even as control is lost.

This book will distinguish **delay** from **control**.

Delay suppresses deviation without restoring the relationships that make stability possible.

Control preserves invariants across repeated disturbance.

The two are often confused because their short-term effects can look identical. In complex systems, they are not the same achievement. I will refer to this difference as **the Delay-Control distinction**.

Before introducing formal notation, it is worth stating what this mathematics is for: not calculation, but recognition.

Equation 1

These equations are not offered as complete biological models. Each is a minimal mathematical skeleton that isolates one structural feature: invariance, constraint, feedback, accumulation, diffusion, amplification, bifurcation, boundedness, and balance. Not every equation in this book is meant to be fully understood on first reading. Recognition matters more than mastery.

Invariance

This equation defines structure by specifying what must remain unchanged when a system is transformed.

What Structure means

$$f(Tx) = f(x) \text{for all } T \in G$$

This idea of invariance is not an abstract curiosity. It sits at the foundation of modern mathematics and physics. In the early twentieth century, Emmy Noether showed that every fundamental conservation law in physics arises from an underlying symmetry, from something that remains unchanged under transformation. Momentum, energy, and angular momentum are not separate facts about the world. They are consequences of invariance.

This same principle appears wherever coherence persists: in music, in living systems, and in perception itself. What endures is not material, but structure.

In music, this is why a melody survives transposition; in life, it is why identity persists even as components turn over.

A transformation T acts on a state x. If f is unchanged for every T in a family G (often a symmetry group), then f is invariant under that family. Invariance is one of the cleanest mathematical signatures of structure: it specifies what can change while something essential remains fixed.

At its most fundamental level, mathematics begins with a simple question:
What remains the same when something changes?

This equation expresses that question formally. A transformation T acts on a system, altering its appearance, position, or representation. Yet the function f remains unchanged. Something essential has been preserved.

This is **invariance**.

Invariance is the mathematical definition of structure. A system has structure if there exists at least one transformation under which it does not change. Without invariance, there is no pattern. Without pattern, there is nothing to recognise, nothing to predict, and nothing to regulate.

Long before mathematics was written symbolically, invariance was observed intuitively. The seasons return despite variation in weather. A melody remains recognisable when played in a different key. A face is recognised across changes in expression and lighting. These recognitions are not trivial. They depend on the detection of what does not vary.

Painters have always worked with invariance, long before mathematics gave it a name.

A face painted by Rembrandt remains recognisable despite profound transformation. Age deepens lines. Light shifts across skin. Expression changes from alertness to fatigue, from defiance to resignation. Yet identity persists. The recognisability of the face does not depend on surface detail. It depends on preserved relationships: proportions, spatial alignment, tension between features. What the eye detects is not colour or texture, but structure.

Rembrandt understood this intuitively. His late portraits are not accurate in the photographic sense. Features are exaggerated, shadows collapse into darkness, contours dissolve. And yet these faces feel more real, not less. The painting survives distortion because the invariants are preserved. When they are violated, the image fails immediately, even if every detail is finely rendered.

Leonardo da Vinci approached the same problem from another direction. His obsession with proportion, perspective, and transformation was not aesthetic indulgence; it was structural inquiry. A figure viewed from different angles remains the same figure only if certain ratios are preserved. Leonardo's notebooks are filled with rotations, projections, and dissections of the human form, not to catalogue parts, but to understand what may change without loss of identity.

Perspective itself is a mathematical transformation. The object shrinks, stretches, or skews as viewpoint changes, yet remains recognisable because relational structure is conserved. Leonardo grasped this centuries before formal group theory. His insight was not that vision obeys mathematics, but that recognition depends on invariance under transformation.

Michelangelo pressed this logic to its limits. His figures are often anatomically impossible: torsos twisted beyond comfort, limbs exaggerated, musculature amplified to excess. Yet they remain coherent. Constraint is not abandoned; it is intensified. Identity survives deformation because the underlying structure is preserved. The body is stretched, but not broken.

This is why a poorly drawn figure feels immediately wrong, even to an untrained observer. The error is not stylistic. It is structural. Invariance has been violated. No amount of surface detail can compensate for that loss.

None of this requires calculation. No one measures angles or ratios while standing before a painting. Recognition is immediate. Error is felt before it is named. Mathematics enters only later, to explain what perception already knows.

This is the same logic that governs music. A melody survives transposition because interval relations are preserved. It governs faces, bodies, and voices. And it governs living systems, whose components change continuously while identity persists.

Mathematics does not invent invariance; it names and formalises the conditions under which recognition remains possible.

In physics, this principle explains why conservation laws exist at all: what is conserved reflects what remains invariant. Invariance under rotation implies conservation of angular momentum. Emmy Noether's insight was that these were not separate facts, but expressions of a single principle. What is conserved reflects what is invariant.

The same logic applies beyond physics. Musical harmony depends on invariance of interval rather than absolute pitch. A melody survives transposition because the relationships between notes remain unchanged. The sound varies. The structure does not.

Life, too, depends on invariance. Cells turn over their components constantly. Molecules are replaced. Energy flows through the system. Yet identity persists. What endures is not material, but organised relationship. Structure is preserved even as substance changes.

This is why life cannot be reduced to chemistry alone. Chemistry describes what is present. Structure describes what is preserved. Without invariance, a system dissolves into transient reactions with no memory.

Invariance also explains constraint. A system governed by structure cannot evolve arbitrarily. Transformations are permitted only if they preserve essential relationships. This restriction is not imposed externally. It is intrinsic. Structure defines a space of possible behaviour.

This space is what makes regulation possible. Regulation does not act on every detail. It acts to preserve invariants.

Temperature, pH, ion concentration, rhythm, and timing are maintained within ranges because deviation threatens underlying structure.

Ageing, disease, and instability can now be framed precisely. They are not random failures. They are **progressive violations of invariance**. What was once preserved becomes approximate. What was once rigid becomes flexible. Eventually, what was invariant fails to hold at all.

Seen this way, the equation $f(x) = f(Tx)$ is not abstract. It is the quiet condition of persistence. It explains why some forms endure and others cannot.

Structure is not everything. But without structure, nothing lasts long enough to matter.

This equation establishes the first term of the book's argument. Before regulation, before noise, before time, there must be something that can be preserved.

That something is structure.

Consequence

To recognise structure is to recognise limits. Not everything is possible, and not every change is allowed. These constraints are not obstacles to life. They are its precondition.

Freedom, coherence, and meaning arise only where something endures across change. Invariance is the mathematical name for that endurance.

Everything that follows in this book depends on it.

A further distinction follows from this.

Systems may exhibit improvement without preserving structure. Function can return locally while invariants continue to erode globally. Time may be gained even as control is lost.

This book will distinguish between **delay** and **control**. Delay suppresses deviation without restoring the relationships that make stability possible. Control preserves invariance across repeated disturbance. The two are often confused because their short-term effects can look similar.

In complex systems, they are not the same achievement. Delay can be extensive in the absence of control, and frequently is. Apparent success may coexist with ongoing drift.

This distinction will recur throughout what follows. It explains why repair does not always imply stability, why intervention does not always imply cure, and why progress can be real without being permanent.

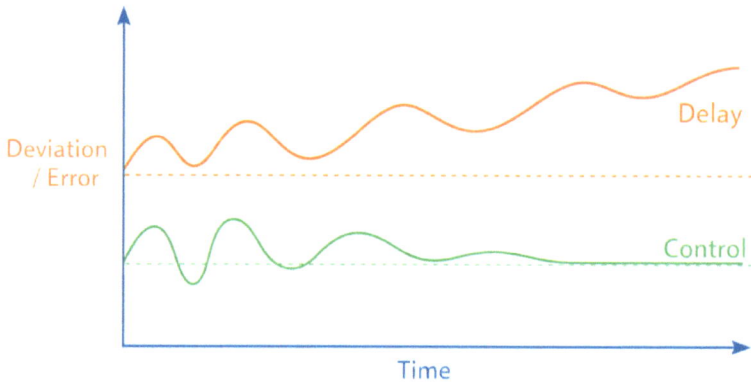

Figure 1. Delay versus control.

Short-term suppression of deviation (delay) can occur without preservation of underlying structure. Control restores invariants and maintains stability under repeated disturbance. The two may appear similar initially but diverge over time.

Interlude 1: High Table

At Oxford, dinner at High Table followed a structure so precise it felt almost ceremonial.

We assembled in order and processed into the hall as the students stood.
The Principal struck the gavel; the Latin grace was said. Only then did we sit, and conversation began.
When the High Table had finished dining, the gavel sounded again and we processed back to the Senior Common Room.

I dined often with a physicist who took these rituals lightly, and the ideas within them seriously. **Ian Shipsey** was internationally known, the head of physics, and entirely uninterested in behaving as if any of that required distance.

We talked through dinner. About medicine, about physics, about the strange pleasure of finding the same questions appearing in different languages. He told me stories about his family, how his ancestors had reached Britain by ship from Ireland, not westward to America as most had done. He liked the idea that small decisions echoed across generations.

His wife, also a professor, would occasionally interrupt to suggest that I might want to eat. Ian would smile and wave her off, saying that talking was the only reason he came to dinner at all.

The food itself was unreliable. We shared a mutual irritation at the random appearance of rocket on plates where it had no clear purpose. We were more forgiving of the pudding, which never disappointed and required no justification.

Outside the hall, he introduced me to particle accelerators at Oxford, not as monuments, but as instruments. He was generous with his attention. When I mentioned ideas about quantum systems, he listened carefully and then introduced me to a colleague who could take them seriously.

I realised later that I had learned something essential in those exchanges, though not what I thought at the time. Ian never separated thinking from building. Questions mattered, but only insofar as they could be met with the right apparatus.

The only time I saw him without a suit was in the Marks & Spencer near the college. We laughed at the coincidence, standing between shelves of ordinary things, and agreed we should go out for dinner somewhere without ritual.

I texted him a few days later to plan it.

He did not reply.

After a day or two, an email arrived instead. He had died suddenly, days earlier.

The Christmas card did not come that year. Nor the next.

The hall still filled. The gavel still fell. The pudding was still good.

But the conversation had ended, mid-sentence.

It struck me then that structure does not announce its failure. It simply continues without what once animated it. The rhythm remains. The signal does not.

Ian Shipsey, Professor and former Head of the Department of Physics, University of Oxford

Chapter 2

Form before Sound

The Mathematics of Music

Music is often described as organised sound. This description is accurate, but incomplete. Sound is the medium through which music becomes perceptible, not the source of its coherence. Long before a note is heard, its form already exists. The mathematics of music precedes its realisation in sound.

A sounding body does not vibrate freely. A string fixed at both ends can oscillate only in particular ways. An air column resonates at specific frequencies determined by its length and shape. These constraints generate a harmonic series, a structured set of relationships that cannot be altered by intention or taste.

This pattern, known as the harmonic series, was recognised as early as Pythagoras and remains one of the simplest demonstrations of how number can structure perception. The ear does not invent consonance and dissonance. It recognises them.

In music, **structure is constraint realised in time**. Certain sounds may follow others, and certain progressions are forbidden. This priority of form over perception gives music its peculiar authority. A note slightly out of tune is immediately perceived as wrong, even by an untrained listener. A rhythm that drifts loses coherence. These judgements are not matters of convention. They are responses to violations of mathematical expectation.

Historically, music was among the first domains in which the abstract power of number was recognised. The discovery that simple numerical ratios underlie musical intervals revealed that mathematics governs quality as well as quantity. Sound became evidence that number can structure human experience. Music therefore occupies a unique position among human activities. It allows mathematics to be heard.

This obedience to structure is made explicit in the discipline of composition. Musical forms rely on repetition, variation, symmetry, and controlled deviation. A theme is introduced, transformed, and returned. Tension arises through departure from expectation and is resolved through its restoration. These processes are intelligible because they are structured. Emotion emerges not in spite of form, but because of it.

In the work of **Johann Sebastian Bach**, musical structure is laid bare. *The Well-Tempered Clavier* systematically traverses all major and minor keys, treating tonal space as a coherent mathematical domain. Each fugue develops an entire movement from a single subject through inversion, augmentation, diminution, and imitation. Complexity arises not from accumulation, but from transformation. Once the initial conditions are established, the remainder follows with near inevitability.

An even starker demonstration appears in the *Goldberg Variations*. Thirty variations unfold over an unchanging harmonic structure. Every third variation is a canon, each at an increasing interval. Difference emerges from constraint. Sound follows structure with exceptional fidelity.

In **Wolfgang Amadeus Mozart**, mathematical logic is absorbed so completely that it becomes almost invisible. His music exhibits balance, proportion, and economy across every genre. Themes are symmetrical. Harmonic tension is introduced and released at precise moments. Surprise is permitted, but never allowed to undermine coherence. What appears effortless is in fact highly constrained.

This concealment of structure reaches a peak in the finale of *Symphony No. 41*, where multiple independent themes are combined simultaneously. Each theme is simple. Their power lies in their compatibility. Counterpoint here is not decorative, but architectural. Order is preserved because the mathematics is exact.

The danger of such fluency is not that structure disappears, but that it becomes inaudible to those who mistake effort for depth.

Popular culture often mistakes this fluency for chaos. In *Amadeus*, Mozart's facility is portrayed as incomprehensible excess, music spilling out faster than it can be written down. The drama lies not in Mozart's abundance, but in Salieri's failure to hear the structure that makes such speed possible. What appears unfiltered is, in fact, fully constrained.

I was reminded of this unexpectedly, in a context far removed from music.

After presenting a framework that argued for an expanded set of cancer hallmarks, explicitly including dedifferentiation and transdifferentiation, epigenetic dysregulation, altered microbiome influence, and altered neuronal signalling, a senior colleague responded not by disputing any individual mechanism, but by noting something else entirely.

He remarked that the work had forced him to recognise how narrowly he had been listening: attending to mutations, pathways, or cell types in isolation, while missing the symphony they formed together.

The comment was not intended as praise. It was diagnostic.

Complexity had accumulated to the point where catalogue was no longer enough. Only form could restore coherence.

This is Mozart's lesson, translated. When structure is sufficiently internalised, it no longer announces itself. It permits many voices at once without losing control. What sounds excessive to a listener focused on parts is, to one attuned to form, inevitable.

With **Ludwig van Beethoven**, the relationship between form and sound changes. Beethoven inherits the same mathematical language, but subjects it to strain. Structure is no longer concealed. It becomes audible. Short motifs are treated as generative units, expanded, fragmented, displaced, and repeated with insistence.

The opening movement of *Symphony No. 5* grows almost entirely from a four note rhythmic figure. The music does not progress by addition, but by transformation. A minimal input generates a complex trajectory. The movement behaves like a dynamical system unfolding from a single initial condition.

This logic extends across unprecedented scale in the *Symphony No. 9*. The final choral theme is deliberately simple. Its strength lies in its malleability. It can be fragmented, layered, inverted, and expanded without losing identity. The long delay before its appearance creates expectation through controlled tension. When the theme finally emerges, it feels inevitable rather than surprising.

These structural developments acquire deeper significance in light of Beethoven's progressive deafness. By the final decade of his life, he could no longer hear the sounds he was composing. Yet many of his most ambitious works belong to this period. Deprived of auditory feedback, Beethoven relied increasingly on internal representation and formal constraint. Composition became an act of mental construction governed by proportion, symmetry, and transformation rather than sensory confirmation.

This is not merely biographical curiosity. It reveals something fundamental. Musical coherence does not depend on sound itself. Sound is the vehicle, not the source. The organising principles of music exist prior to their physical expression. When perception fails, structure remains.

Music also clarifies how expectation governs perception. A listener does not experience each moment in isolation. The nervous system continuously predicts what should come next. Rhythm partitions time into ratios. Harmony establishes probabilistic expectation. Form governs anticipation across longer spans. Satisfaction arises not from novelty alone, but from confirmation and return.

This predictive structure explains why musical errors are immediately perceptible. A wrong note violates expectation. A mistimed entrance disrupts coherence. Musical understanding is therefore an exercise in error detection. Mathematics governs not only composition, but perception.

Performance introduces controlled variability. No two performances are identical, yet the identity of the work persists. Timing fluctuates. Dynamics shift. Expression varies. Structure remains. Robust mathematical organisation tolerates deviation without collapse. Too little variation produces sterility. Too much destroys coherence. Musical performance operates within a narrow corridor between rigidity and randomness.

This balance anticipates a central theme of later chapters. Living systems must preserve structure while tolerating noise. Too much error leads to dysfunction. Too little flexibility leads to brittleness. Music offers a transparent demonstration of this principle in action.

Music therefore represents a limiting case. A note slightly out of tune is recognised instantly, even by someone who cannot name it. Its constraints are explicit. Its feedback is immediate. Its failures are audible. It shows what mathematical regulation looks like when obedience is tight and instability is immediately punished.

Here, form comes before sound.

In the chapters that follow, the same mathematical principles will reappear under increasingly demanding conditions. In living systems, regulation must persist amid damage, uncertainty, and noise. In ageing, precision erodes gradually. In cancer, control fails catastrophically.

Interpretation

Music teaches that meaning does not begin with expression. It begins with form. Sound gives form a voice, but it does not create it.

What we hear as beauty or coherence is the consequence of relationships organised in time. When those relationships are strong, expression survives strain. When they weaken, no amount of intensity can restore meaning.

Form endures. Sound passes.

Equation 2

The Harmonic Series
A foundational pattern in waves and resonance

Why Music exists before Sound

This equation shows that a constrained system can persist only in specific, permitted modes.

This relationship appears wherever vibration is constrained, in strings, air columns, membranes, and later in quantum mechanics, because it reflects boundary conditions rather than musical convention.

$$f_n \approx n f_1, \, n = 1, 2, 3, \ldots$$

For an ideal string fixed at both ends (or an ideal resonant column with standard boundary conditions), oscillation is permitted only at integer multiples of a fundamental frequency. Real instruments depart slightly from exact integer ratios because of stiffness, damping, and geometric end effects, so the harmonic relation is typically approximate rather than exact.

The principle, however, is unchanged. No matter how the system is excited, only these modes persist. Structure does not describe what is heard; it determines what can endure.

This is not a convention. It is a constraint.

The harmonic series does not describe music as we hear it. It describes the **space of sound that is possible** before any sound is made. Music exists because this space is ordered.

A string plucked at random does not produce random sound. It produces a structured spectrum: some frequencies persist, others cannot. Harmony arises not from choice, but from obedience to constraint.

This is part of why music remains intelligible across cultures and history. Styles vary. Instruments differ. But the underlying mathematics does not change. Intervals that sound consonant do so because their frequency ratios are simple. Complexity emerges from repetition, not from abandoning structure.

The harmonic series demonstrates a crucial principle: **form precedes experience**.

Before a listener hears anything, before a composer writes a note, the structure already exists. Sound realises mathematics; it does not invent it.

This explains a long-standing mystery. Why does music feel inevitable when it is well formed? Why do resolutions feel earned rather than arbitrary? Because the system is moving through a constrained space. Certain paths lead back to stability. Others do not.

Musical expectation is therefore mathematical. A listener anticipates resolution not by memory alone, but because the structure biases what can plausibly follow. Surprise is meaningful only when it occurs within these limits.

This is why wrong notes are immediately recognisable. They violate structure, not taste. The nervous system registers the violation before conscious judgement intervenes.

The same logic explains transposition. A melody remains recognisable when shifted to a different key because the ratios between notes are preserved. Absolute frequencies change. Relationships do not. Invariance returns.

This property reveals why music is the cleanest example of mathematical order available to human experience. Feedback is immediate. Error is audible. Constraint is strict. There is little tolerance for sustained deviation.

For this reason, music serves as the ideal starting point for understanding life. It shows what mathematics looks like when regulation is perfect and noise is minimal. Every deviation is corrected instantly or exposed.

Life will relax these conditions. Noise will intrude. Correction will be delayed. Structure will be preserved approximately rather than exactly. But the logic will remain.

The harmonic series is not merely about sound. It is about **permitted behaviour**. It shows how richness emerges not from freedom, but from restriction.

Without this equation, music would be noise.
Without constraint, form would dissolve.

Consequence

Music teaches that meaning arises when variation occurs within limits. Absolute freedom produces randomness. Absolute rigidity produces monotony. Harmony exists between these extremes.

This balance is not aesthetic preference. It is mathematical necessity.

What music makes audible; life will make fragile.

Interlude 2: Silence, Error, and Expectation

Music does not exist only in sound. It exists equally in silence.

Between notes, between phrases, and between movements, silence carries structure. It is not the absence of music, but the space in which expectation forms. A rest placed correctly is as consequential as a note played at the right pitch. A rest placed incorrectly collapses coherence just as surely as a wrong note.

This sensitivity to absence reveals something essential. Musical understanding depends not on what is heard, but on what is *anticipated*. The listener continuously predicts what should come next. When the prediction is met, coherence is reinforced. When it is violated slightly, tension arises. When it is violated excessively, meaning dissolves.

Expectation is therefore the quiet engine of musical experience.

This process is mathematical in nature. Rhythm divides time into ratios. Meter establishes periodicity. Harmonic progression constrains what transitions are likely, permitted, or forbidden. These structures do not dictate every detail, but they define a space of expectation. Even unfamiliar music remains intelligible if its internal logic is preserved.

This is why musical mistakes are immediately perceptible. A wrong note is not merely different. It violates a prediction generated by form. The nervous system registers this discrepancy automatically. Musical understanding, at its core, is an exercise in error detection.

Silence sharpens this process. When sound stops, expectation does not. In fact, it intensifies. The listener continues the music internally, projecting forward, awaiting resolution. Silence reveals that music exists as a structure held in the mind, not merely as vibration in air.

This internal continuation is not imagination in the casual sense. It is inference. The brain maintains a model of the music and updates it as sound arrives. When sound ceases, the model persists. When sound resumes, the model is either confirmed or corrected.

This interplay between prediction and correction is exacting. Too much predictability produces boredom. Too much surprise produces confusion. Meaning arises only within a narrow corridor between these extremes. Music that survives does so because it occupies this region reliably.

Silence exposes another crucial feature. Correction is asymmetric. When expectation is violated, the listener adjusts. But the adjustment is never perfect. Slight uncertainty remains. Over time, repeated small surprises change how future passages are heard. Experience accumulates.

Even in music, then, there is drift.

In a perfectly obedient system, drift is limited. The score constrains deviation tightly. Performers may vary timing or dynamics, but structure remains dominant. Feedback is immediate. Errors are punished audibly. The system tolerates little ambiguity.

Life will not be so strict.

Yet the principle carries forward. Living systems also depend on expectation. Cells anticipate nutrient availability. Organisms anticipate threat. Nervous systems anticipate sensory input. In each case, prediction precedes response. Correction follows deviation.

The difference is that life must operate continuously, without rests.

There is no silence in biology. Noise persists. Signals overlap. Corrections occur while new disturbances arrive. Regulation must operate not in discrete moments, but in ongoing flow. Expectation never resets cleanly. Error accumulates.

Music allows us to isolate the logic of expectation because it is disciplined by form. Silence gives us a glimpse of how prediction works when structure is clear and noise is minimal. Life will require the same mathematics under far less forgiving conditions.

This is why music is such a powerful starting point. It strips regulation to its essentials. It shows what happens when expectation is precise, error is obvious, and correction is immediate.

The chapters that follow will move away from this clarity. Regulation will become probabilistic. Correction will be delayed. Silence will disappear. Expectation will be burdened by uncertainty.

But the underlying logic will not change. What music makes audible, life makes fragile.

Interlude 3: Neural Regulation Under Constraint

Neurons are often described as information processors. This description is convenient, but misleading. A neuron does not primarily compute in the way a digital system does. It regulates.

At every moment, neural systems operate under uncertainty. Sensory input is incomplete, delayed, and noisy. Motor output is imprecise. Internal states fluctuate. Yet behaviour remains coherent. This coherence is not achieved by exact calculation, but by continual prediction and correction.

Neural circuits generate expectations about incoming signals before those signals arrive. Sensory input is compared against these expectations, and what is transmitted forward is not raw data, but deviation. The nervous system is organised around error.

This architecture is not incidental. It is the only viable strategy for regulation under noise and time. Waiting for certainty would be fatal. Acting without prediction would be chaotic. Neurons therefore anticipate, adjust, and update continuously, preserving coherence **without requiring exact precision**.

From a mathematical perspective, neural systems exemplify regulated instability. Activity is never static. Baseline firing fluctuates. Synaptic strengths drift. Networks oscillate. Yet these fluctuations remain bounded for as long as regulation holds.

Plasticity, often celebrated as the basis of learning, is also a source of vulnerability. Synaptic modification allows adaptation, but it also introduces drift. Each update reflects past conditions more than future ones. Over time, accumulated misalignment subtly degrades predictive accuracy. Correction continues, but it arrives later and with less authority.

Ageing in neural systems is therefore not best understood as loss of neurons alone. It is loss of precision in prediction. Signals remain, but their timing degrades. Thresholds blur. Noise is tolerated less effectively. The system still functions, but closer to its boundaries.

When regulation fails more abruptly, neural behaviour changes regime. Epileptic discharges, for example, are not excessive activity in the ordinary sense. They are loss of constraint. Local amplification overwhelms inhibitory control. Feedback reverses sign. The system enters a new dynamical state.

The distinction is easiest to see outside the clinic.

A brief clinical vignette makes this visible.

I once encountered a young woman who developed a generalised seizure in a crowded restaurant. Conversation fractured instantly. Chairs moved back. Voices rose. The social system destabilised faster than the neural one.

The seizure itself was brief. What persisted was loss of constraint, synchronous activity overwhelming inhibition, spreading because nothing opposed it.

Nothing new needed to be introduced to restore coherence. The intervention was subtractive rather than additive: space cleared, stimuli reduced, time allowed. As inhibitory control reasserted itself, the system returned to baseline.

What struck me was not the drama of the event, but its structure. A stable system had crossed a threshold. Behaviour changed regime. When constraint returned, coherence followed rapidly, without repair or recovery.

The same misunderstanding appears in teaching.

When epilepsy is demonstrated in the lecture hall, it is often received as excess, too much firing, too much excitation. Yet the patients themselves describe something else: not intensity, but loss of boundary. Not chaos, but synchronisation without control.

What fails is not the neuron.

What fails is regulation.

This logic mirrors what has already been described in cancer. The substrate differs. The mathematics does not.

Neurosurgical disease makes this explicit. Tumours distort connectivity rather than simply destroying tissue. Oedema alters timing. Compression changes thresholds. Function is often lost not because neurons are absent, but because regulation has become incoherent. Relief of pressure can restore function rapidly, revealing that structure persisted even when control was compromised.

This distinction matters clinically and conceptually. It shows that coherence depends less on material integrity than on preserved regulation. Neurons can survive substantial insult and still function if coupling and feedback remain intact. Conversely, intact tissue can fail catastrophically when regulation collapses.

The nervous system therefore does not contradict the book's argument. It exemplifies it.

Neural life persists not through precision, but through continual correction under uncertainty. Meaning, perception, and action arise not because the system is exact, but because it remains bounded long enough for patterns to be recognised.

Neurons do not escape mathematics.
They live inside it.

A minimal mathematical example: Neural threshold and regime change

The behaviour of a neuron can be captured, at its simplest, by a single differential equation:

$$\frac{dV}{dt} = -\frac{1}{\tau}(V - V_{rest}) + I(t)$$

This is a leaky integrator model, one of the simplest and most widely used descriptions of neuronal dynamics.
Here, V represents membrane potential, V_{rest} the resting state, τ a time constant, and $I(t)$ incoming input.

This equation describes regulation. Left alone, the system relaxes back toward its baseline. Perturbations decay. Stability is preserved.

Yet this stability is conditional.

When input pushes V beyond a threshold imposed on this otherwise linear dynamics, the system undergoes a qualitative change. A spike is generated. The neuron resets. Behaviour is no longer proportional to input. A boundary has been crossed.

This is not failure. It is controlled instability.

Under normal conditions, thresholds are enforced, timing is regulated, and activity remains bounded. Under pathological conditions, loss of inhibition, altered thresholds, delayed correction, the same equation produces runaway firing, synchronisation, or collapse into abnormal rhythms.

Nothing in the equation has changed.

The regime has.

This is why neural disease is not best understood as excess activity, but as loss of constraint. And it is why restoration of regulation, rather than suppression of activity alone, often restores function.

The mathematics is simple.
The consequences are not.

Chapter 3

Regulation before life

Mathematics in living systems

If music shows what mathematics looks like under conditions of strict obedience, life shows what happens when those conditions are relaxed. Living systems do not follow equations with the precision of a vibrating string. They operate in environments that are noisy, uncertain, and continually changing. Yet they remain coherent. This coherence is not accidental. It is regulated.

Life is not chemistry alone; it is chemistry under regulation. By regulation we mean the set of processes through which a system detects deviation from expected behaviour and acts to counteract it. Regulation is what allows organisation to persist over time despite continual disturbance. This capacity is mathematical in nature. Life is not exempt from mathematics. It is sustained by it.

At the most basic level, living systems exist far from equilibrium. Left to themselves, physical systems tend towards disorder. Living systems resist this tendency by continuously exchanging energy and matter with their environment. This resistance is not static. It is dynamic. Stability is achieved not by remaining unchanged, but by continually correcting deviation.

In classical Tamil grammar, pronunciation is not measured numerically. It is measured in time.

A syllable carries a *māttirai*, a unit of duration rather than quantity. Vowels are not simply short or long; they occupy permitted temporal spans. An *arai māttirai* is half a unit of time. A *kāl māttirai* is a quarter. These are not fractions in the abstract sense. They are constraints on rhythm: how long a sound may exist before it breaks the structure of speech.

No one calculates these durations. They are learned through repetition, correction, and ear. The mathematics is enforced by coherence. A vowel held too long or released too quickly is not "incorrect" by rule; it simply feels wrong. Meaning collapses when timing drifts.

Classical Chinese thought encodes quantity in a similar way. Proportion precedes number. Concepts like *bàn* (half) or *fēn* (part) are relational rather than absolute. In medicine, music, and poetry, balance matters more than count. Measure is contextual, anchored to harmony rather than exact enumeration.

Latin preserves traces of the same logic. Words like *dimidium*, *tertia pars*, and *quarta* refer not to abstract decimals, but to sanctioned divisions of a whole, of land, of time, of obligation. These were permissions before they were numbers.

Across these traditions, mathematics appears first as constraint in time and relation, not as symbol or calculation. Structure comes before number. Rhythm comes before measure.

Only later does mathematics detach from lived experience and become autonomous.

In intensive care, stability is maintained not by stillness, but by constant small adjustments, oxygen nudged, fluids titrated, pressure supported, often before deviation becomes visible.

Regulation is mathematically formal. It depends on feedback, thresholds, delays, redundancy, and hierarchy. Negative feedback counteracts deviation and restores balance. Positive feedback amplifies change when transitions are required. Time delays introduce oscillation. Redundancy distributes risk. These are not metaphors borrowed by biology. They are properties of dynamical systems realised in living matter.

Physiology makes this explicit. Heart rate, respiration, hormone release, and neural activity are all governed by regulated oscillations. These rhythms are not imposed from outside. They emerge from interacting feedback loops operating across multiple scales. Circadian rhythms, metabolic cycles, and cell division are governed by the same regulatory logic.

Unlike music, however, life must tolerate error. Molecules collide stochastically. Signals degrade. Components fail. Regulation therefore cannot rely on exact repetition. It must accommodate variability without losing identity. This requirement introduces a fundamental shift. Biological mathematics is probabilistic.

Cells do not respond deterministically to single signals. They integrate multiple inputs, weight them, and respond according to thresholds. Outcomes are not guaranteed, only biased. This stochasticity is not a defect. It is essential. It allows populations of cells to adapt, differentiate, and survive under changing conditions.

This distinction is crucial. Music cannot survive drift. A rhythm that slowly accelerates or decelerates loses coherence. Life, by contrast, is built to absorb drift. Regulation compensates continuously, recalibrating internal variables to maintain function. Health is not perfect order. It is controlled variability.

Regulation must operate across many scales simultaneously. Molecular reactions unfold in microseconds. Cellular processes span minutes or hours. Physiological rhythms extend over days. Development and ageing unfold over decades. These scales are coupled. A disturbance at one level can propagate to others, sometimes amplifying, sometimes dissipating.

Effective regulation therefore requires hierarchy. Fast processes are stabilised locally. Slower processes provide context and constraint. This hierarchical organisation allows living systems to remain stable without becoming rigid. Local fluctuations are tolerated so long as global organisation persists.

A clinical scene makes this visible.

During my rotation in paediatrics, I encountered a child with a rare metabolic condition. The details mattered clinically, but what stayed with me was not the diagnosis.

The child required constant care. Feeding, monitoring, positioning, nothing could be skipped without consequence. The parents moved around the cot with an efficiency that came only from repetition. When one stepped away, the other adjusted without comment. There was no visible hierarchy, only continuity.

They were not optimistic. They were organised.

Around the same time, I watched another family fragment. Both parents were physicians. Their child had autism. The demands were different, but the strain was constant. Appointments multiplied. Sleep disappeared. Decisions accumulated.

The diagnosis itself was not the source of instability; the system failed because sustained, unrelieved demand gradually dissolved coordination.
Gradually, they stopped coordinating.

Care continued, but no longer coherently. Responsibilities were duplicated or dropped. Information failed to transfer. What one parent assumed the other would manage often went undone.

This is regulation at the level of coupling: local competence without global coherence.

No one intended this outcome. There was no singular failure. The system simply lost its ability to regulate itself under sustained load.

In medicine we often focus on the burden placed on individuals. Less often do we notice when the problem is not the size of the disturbance, but the loss of coupling between parts.

Both families loved their children.

Only one system remained stable.

This loss of coordination is not dramatic. It does not require failure at every level. Regulation can persist locally while coherence is lost globally.

Redundancy is central to this robustness. Cells die, proteins misfold, signals misfire. Yet organisms persist. Mathematical organisation distributes risk. No single failure is decisive. Regulation succeeds precisely because it does not demand precision everywhere at once.

Regulation is often described as control, but control requires information. Living systems must detect deviation, estimate its significance, and respond appropriately. This process carries energetic cost. Regulation is never free.

There is therefore a fundamental trade-off between sensitivity and stability. Systems that respond too aggressively to noise become unstable. Systems that respond too slowly fail to correct error. Effective regulation lies between these extremes. Living systems occupy this narrow region by maintaining margins of safety. Excess capacity and apparent inefficiency are not design flaws. They are features of robust control.

The same logic applies at the cellular level. Cells estimate nutrient availability, stress, and damage, adjusting gene expression accordingly. These adjustments are probabilistic rather than exact. Regulation succeeds not by eliminating uncertainty, but by managing it.

This emphasis on prediction marks a critical transition. Living systems do not merely react. They anticipate. At every scale, regulation depends on expectation, error detection, and correction. When prediction fails repeatedly, control weakens. Variability grows. Instability follows.

Life therefore occupies a narrow mathematical regime. Too much order prevents adaptation. Too much variability destroys coherence. Living systems operate close to the boundary between stability and instability, where responsiveness is high but control is still possible.

This proximity to breakdown is both a strength and a vulnerability. It enables rapid adaptation to changing environments, but it also means that regulation can be lost. When compensation becomes imperfect, error accumulates.

Ageing and cancer arise within this same mathematical landscape. Ageing reflects the gradual loss of regulatory precision. Cancer reflects runaway instability when control mechanisms fail. Neither is foreign to life. Both emerge from the same mathematics that ordinarily sustains it.

Life does not violate physical law. It negotiates with it continuously.

Music showed what mathematical order looks like under ideal conditions, where feedback is immediate and deviation is punished. Life shows what mathematical order looks like under pressure, where noise is tolerated, prediction is essential, and correction is continual.

In the next chapter, this pressure will be shown to accumulate. Regulation is not eternal. Precision erodes. Compensation becomes incomplete. The mathematics of life begins to drift.

Ageing is not the opposite of life. It is what happens when regulation slowly loses accuracy.

Modern medicine has become extraordinarily good at delaying failure. It is far less good at restoring control. When these two achievements are treated as equivalent, expectations drift away from what complex systems can actually sustain.

Interpretation

Life is sustained not by precision, but by correction. Error is not an anomaly to be eliminated, but a condition to be managed. Health lies not in stillness, but in the continual restoration of balance.

What distinguishes living systems is their capacity to remain coherent while never being exact. They endure by anticipating change, absorbing fluctuation, and correcting drift before it becomes collapse.

Stability, in life as in mathematics, is achieved neither by rigid control nor by surrender to chance, but by maintaining structure while allowing movement. Life endures not by resisting mathematics, but by living within its limits.

Equation 3

Negative Feedback

How regulation sustains life

This is the simplest equation of homeostasis, the logic behind thermostats, autopilots, and physiological regulation: the further a system moves from its preferred state, the stronger the pull back.

$$\frac{dx}{dt} = -k(x - x_0), \qquad k > 0$$

When the system deviates from its preferred state x_0, change is redirected back toward it. The larger the deviation, the stronger the correction. In the body, this is why temperature, glucose, and neural activity fluctuate continuously yet remain bounded rather than running away.

This is negative feedback.

Negative feedback does not eliminate change. It makes change reversible. Deviations occur, but they are countered. Stability emerges not from stillness, but from continual correction.

This distinction matters. A system without feedback may appear stable if undisturbed, but it cannot respond.

This explains why living systems exist far from equilibrium. Equilibrium is not health. It is death. Regulation maintains distance from equilibrium by continuously expending energy to oppose drift.

Negative feedback also introduces time. Correction is not instantaneous. Delays generate oscillation. Overcorrection produces instability. Effective regulation depends on timing as much as magnitude.

This is why regulation must be tuned. If correction is too weak, deviation persists. If it is too strong, the system overshoots. Stability lies between sluggishness and aggression.

This balance is mathematical, not metaphorical.

The equation above is linear and idealised. Real biological regulation is nonlinear, distributed, and noisy. Yet the principle remains. Correction opposes deviation.

Without this mechanism, life cannot exist.

Negative feedback also explains robustness. Because correction responds to deviation rather than cause, the system does not need to identify every disturbance. It simply needs to detect that something has changed. Regulation succeeds even when the source of noise is unknown.

This property makes life possible under uncertainty.

However, regulation is not perfect. Feedback loops degrade. Sensors drift. Actuators weaken. Correction becomes approximate. The same equation continues to operate, but its parameters change.

The failure of regulation is not sudden. It is gradual erosion of feedback effectiveness. Deviation is still opposed, but less precisely. Overshoot increases. Delays lengthen. Oscillations widen.

This equation therefore marks both the birth of life and the beginning of its vulnerability.

Without negative feedback, there is no persistence.
With imperfect feedback, there is drift.

Consequence

Life is not maintained by equilibrium, but by resistance to it. Health depends not on eliminating change, but on correcting it continually.

Control is not optional. It is the cost of persistence.

As long as feedback dominates, coherence survives. When feedback weakens, instability becomes possible.

Everything that follows in this book describes what happens as this equation loses accuracy over time.

Interlude 4: The body as a prediction machine

Regulation is often imagined as response. Something happens, and the system reacts. This picture is incomplete. Living systems do not wait for disturbance. They anticipate it.

The body is not primarily reactive. It is predictive.

At every scale, living systems generate expectations about the world and about themselves. These expectations guide behaviour before signals fully arrive. Action precedes confirmation. Correction follows error.

This logic is visible in its simplest form in posture. Standing upright requires continuous adjustment. Muscles activate before imbalance becomes perceptible. If the body waited to react, balance would already be lost. Stability depends on prediction.

The same principle governs perception. Sensory systems do not passively record input. They compare incoming signals against internal models. What is experienced is not raw data, but deviation from expectation. Surprise, not sensation, drives attention.

This predictive organisation is mathematical. The system maintains a model of likely states. Incoming information updates that model by reducing error. Regulation succeeds when prediction is accurate enough that correction remains small.

Crucially, prediction is never perfect. Signals are noisy. Environments change. Models lag behind reality. The body therefore operates probabilistically, weighting possibilities rather than selecting certainties.

This behaviour is not philosophical. It is mathematical necessity.

This probabilistic inference is not a defect. It is the only way to function under uncertainty. Deterministic systems fail catastrophically when assumptions break. Predictive systems degrade gracefully.

The metabolic system illustrates this clearly. Energy is allocated in advance of demand. Glucose is stored, mobilised, and redistributed based on expectation rather than immediate need. Hormonal regulation operates on forecasts shaped by history, rhythm, and context.

Immune regulation follows the same logic. Threats are anticipated based on prior exposure. Responses are biased toward likely dangers rather than optimised for each encounter. This strategy trades precision for speed. It preserves coherence at the cost of occasional error.

Even cellular regulation exhibits prediction. Gene expression reflects anticipated conditions rather than current ones. Cells commit to differentiation pathways before outcomes are guaranteed. Life advances by inference, not certainty.

This anticipatory structure explains why living systems are vulnerable to drift. Predictions are shaped by the past. As time passes, accumulated error subtly distorts expectation. Regulation continues to operate, but it operates on increasingly outdated models.

Ageing begins here.

Not with damage, but with misalignment.

The body continues to predict, but predictions lose accuracy. Correction still occurs, but it arrives late or overshoots. Rhythms fall slightly out of synchrony. Thresholds blur. What was once tight regulation becomes approximate.

Importantly, the system does not recognise this as failure. Prediction does not signal its own degradation. From the inside, function feels continuous. This is why ageing is rarely perceived until margins are exceeded.

The same predictive logic explains fatigue, pain, and stress. These are not merely responses to damage. They are signals that uncertainty has increased, that error correction is becoming costly, that regulation is approaching its limits.

Prediction also explains why living systems are meaning-making. To predict is to assume continuity. To anticipate is to value persistence. The body behaves as if the future matters because regulation requires it to.

A system that did not predict could not regulate. A system that did not regulate could not persist. A system that could not persist would never experience anything at all.

The next chapter will show what happens when prediction slowly loses fidelity. Regulation remains, but its models drift. Correction continues, but with diminishing precision.

Ageing is not the failure of prediction.
It is the cost of relying on it for too long.

Interlude 5: When the Infinite Presses Back (On Scale and Human Limits)

At one moment in his life, Blaise Pascal was left suspended above empty space. The carriage in which he travelled came to rest with its forward weight hanging beyond the bridge, the horses gone, the ground no longer beneath him but somewhere below. Nothing moved. There was no fall, only the knowledge that there could have been one. The event passed, but the configuration remained: support behind, nothing ahead, the body held at the boundary between continuity and collapse.

Pascal had already been working at that boundary in mathematics. He had given form to uncertainty, learned how chance accumulates, how outcomes become predictable without becoming certain. He understood that infinity need not be reached to exert force. From that point on, the void was no longer an abstract absence. It had scale, direction, and immediacy. Mathematics could describe it. The nervous system could not ignore it.

Isaac Newton encountered instability from the opposite side. Where Pascal balanced above an opening, Newton tightened his world until openings disappeared. His mathematics disciplined motion, forced change into limits, reduced flux to derivation. Approximation was tolerated only if it could be controlled, error admitted only if it could be bounded. Stability was achieved by constraint, but constraint accumulated. Precision narrowed the domain until flexibility was lost and the system became brittle.

Georg Cantor would later remove the boundary altogether. Infinity was no longer a horizon or a threat; it became an object of internal structure. He ordered it, counted it, stratified it. What had once been an edge became an interior space. Mathematics crossed a threshold, exact, irreversible, and correct. But the human system carrying that transition did not remain unchanged. Precision did not buffer the effect; it amplified it.

These are not failures of reason. They are encounters with scale. Mathematics can enter regimes where the assumptions of biological regulation no longer apply. It can stabilise infinity, formalise uncertainty, and control motion arbitrarily close to singularity. The body cannot follow without consequence. Fear appears not because something is wrong, but because nothing is bounded.

What emerges, in such moments, is not error but a change of regime. Behaviour shifts abruptly. Language thins. Avoidance replaces exploration. The world has not become more dangerous; the internal model has lost its damping. Prediction outruns correction.

The mathematics remains true. It always does.
What changes is the cost of holding it.

Chapter 4

Drift over Time

Mathematics and Ageing

Regulation sustains life, but it does not halt time. Every regulatory process operates with finite precision. Signals are delayed. Corrections are approximate. Repair is incomplete. Over time, these small imperfections accumulate. Ageing begins here.

Ageing is often described as damage, decline, or wear. These descriptions capture effects, not cause. At its core, ageing is best understood mathematically. It is the gradual loss of accuracy in regulatory control. What once held behaviour within narrow bounds begins to wander.

In the short term, these residues are negligible. Over longer timescales, they are not. Drift is the inevitable consequence of imperfect regulation applied repeatedly.

Living systems depend on prediction. They anticipate expected states and act when reality deviates from expectation. Prediction, however, relies on internal models that must themselves be updated through experience. Over time, these models degrade. Noise infiltrates estimation. Thresholds blur. Feedback weakens.

The result is not immediate dysfunction, but widening variability. Rhythms lose synchrony. Responses become slower or exaggerated. Systems remain functional, but less precise. Ageing does not announce itself through collapse. It reveals itself through dispersion.

This is why recovery stops being clean. A poor night's sleep no longer resolves itself by morning. A minor illness leaves a trace that lingers. Exertion once absorbed without consequence now echoes for days. Nothing dramatic has failed. The system still corrects. It simply does so less exactly, leaving small residues that were once erased.

This distinction separates ageing from disease. Disease involves specific breakdowns. Ageing involves a global loss of regulatory fidelity. The same mechanisms continue to operate, but with increasing error. There is no single lesion to repair, because nothing has failed in isolation.

Biology describes this drift in many languages; the list below is a translation guide, not an argument.

They are often treated as distinct causes. Seen mathematically, they are not causes, but correlated expressions of drift.

Genomic instability does not arise because repair ceases, but because repair is probabilistic. Over time, residual error accumulates. What changes is not the presence of repair, but its precision and coordination across the genome.

Telomere shortening is frequently framed as a counting mechanism. More accurately, it represents a boundary condition. Telomeres define the margin within which replication remains safe. As that margin narrows, division becomes increasingly constrained. Risk rises not because the system has changed purpose, but because tolerance has diminished.

Epigenetic drift offers a particularly clear example. Epigenetic marks regulate when and where genes are expressed. These marks are copied, erased, and rewritten continually. Over time, copying becomes noisier. Signals blur. Cells retain identity, but with less exactness. The genome remains stable. Regulation does not.

Loss of proteostasis follows the same logic. Proteins misfold constantly and are corrected or removed. Ageing does not begin when misfolding appears. It begins when correction becomes slightly less reliable. Aggregates form not because regulation vanishes, but because its error rate increases.

Mitochondrial dysfunction reflects drift in energetic regulation. Energy production continues, but coupling between supply and demand degrades. Variability increases. Efficiency declines. Compensation persists, but at rising cost.

Stem cell exhaustion is not simple depletion. Stem cells remain present, but the regulatory signals governing division, differentiation, and repair lose alignment. Renewal slows because coordination drifts, not because capacity disappears.

Altered intercellular communication completes the picture. Living systems rely on coordination across scales. As timing degrades and thresholds shift, signals arrive late, weak, or distorted. Regulation becomes fragmented. Systems remain alive, but less integrated.

Considered separately, these hallmarks appear heterogeneous. Considered together, they describe a single mathematical phenomenon: the widening of variance as regulatory precision declines.

Ageing is therefore not a catalogue of failures, but a redistribution of probability. Rare events become common. Marginal states become occupied. Correction continues, but less consistently. What was once tightly regulated becomes merely bounded.

This explains why ageing manifests differently across tissues and individuals. Drift does not proceed uniformly. Some regulatory loops degrade faster than others. The result is heterogeneity rather than uniform decline.

It also explains why interventions aimed at individual hallmarks rarely reverse ageing. Repairing one domain does not restore global coordination. What has drifted is not a part, but the alignment between parts.

Ageing is thus a systems-level phenomenon. The hallmarks are symptoms of a deeper mathematical shift: regulation operating with diminishing accuracy under the action of time.

Importantly, drift does not imply instability. A system can remain stable while becoming less precise. It continues to function, but with shrinking margin. Resilience decreases. The cost of perturbation rises.

This narrowing margin defines ageing. Regulation persists, but its effectiveness wanes. Correction continues, but arrives late or incomplete. Control remains, but with less authority.

Ageing is not the opposite of regulation. It is what regulation becomes when time is allowed to act.

In the next chapter, this distinction will sharpen. Drift is gradual and diffuse. Instability is abrupt and localised. When regulatory precision falls below a threshold, behaviour changes regime.

Interpretation

Ageing is not best understood as damage accumulating in parts, but as precision dissolving in control. What erodes is not structure itself, but the ability to keep structure aligned.

Systems do not fail because they stop functioning, but because they function with growing error. Stability can persist long after accuracy is lost.

To age is therefore not simply to decline, but to drift. What matters is not whether regulation remains, but how precisely it continues to operate.

Equation 4

Error Accumulation

Why Ageing is inevitable

This equation shows why small residual errors accumulate over time even when regulation and repair remain active.

In mathematics, this is the simplest form of a **random walk**, a model first studied to understand Brownian motion and now used everywhere from finance to biology.

$$x_{t+1} = x_t + \epsilon_t$$

In ageing, this explains why recovery returns function without fully restoring the original state.

If the errors average to zero, the system need not drift in any particular direction, but its spread still grows with time. If the average is not zero, a directional drift appears.

This is the simplest model of accumulated residual error: each step adds a small term ε_t. When the errors are unbiased ($E[\varepsilon_t] = 0$) and have finite variance, the mean may remain stable while variability increases. When the errors are biased ($E[\varepsilon_t] \neq 0$), systematic drift emerges. The equation appears trivial. It is not.

At each step, the system carries forward its current state and adds a small error term, **the trace of an imperfect correction**.

The error may be positive or negative. It may be random. It may be tiny. What matters is that it is not zero.

Over time, these errors accumulate.

No correction is perfect. Sensors have noise. Actuators have limits. Repairs introduce their own imperfections. Regulation reduces error, but it does not eliminate it entirely. What remains is carried forward.

This is ageing, mathematically.

The system continues to function. Feedback remains active. Stability persists. Yet precision erodes.

This equation explains why ageing is gradual. Each step introduces an error too small to matter on its own. There is no threshold event. No singular failure. Time does the work.

Importantly, error accumulation is neutral with respect to direction. There is no programme pushing the system toward decline. Drift emerges because perfect symmetry cannot be maintained indefinitely.

This reframing dissolves a common misconception. Ageing is often described as wear and tear, or as damage overwhelming repair. In mathematical terms, ageing is more subtle. It is the consequence of integrating noise under continued control.

Repair continues. Regulation operates. Yet each cycle leaves a trace.

This explains why ageing is heterogeneous. Different subsystems accumulate error at different rates. Some remain precise long after others drift. The organism persists by redistributing load, compensating where possible.

But redistribution itself carries cost.

The equation also explains why early life appears robust. When margins are wide, accumulated error remains far from thresholds. Drift is present, but invisible. Only later does accumulated deviation begin to matter.

Ageing is therefore not a failure of regulation, but its shadow.

This equation also introduces irreversibility. While individual errors may cancel, variance increases. The system's past becomes encoded in its present state. Time acquires direction.

Even if regulation were restored perfectly at a later point, the accumulated deviation would remain. The path cannot be retraced exactly. This is why no amount of rest in later life fully recreates the ease of earlier recovery. What was once erased overnight must now be actively managed, and some of it is never quite removed.

This is not pessimism. It is arithmetic.

Ageing is not something that happens to life.
It is what happens when life persists.

Consequence

Ageing does not signal breakdown. It signals endurance. A system that did not accumulate error would not be living. It would be static.

Drift is the price paid for persistence under noise.

Understanding ageing in this way shifts the question. The problem is not how to stop error, but how long regulation can continue to absorb its consequences.

Interlude 6: Time Experienced

Time is not experienced as a sequence of identical moments. It is felt as density.

Some periods pass almost unnoticed. Others seem to thicken, stretching and pressing against attention. A single minute can feel longer than an entire day. These variations are not psychological curiosities. They reflect how regulation interacts with accumulation.

In physical systems, time is a coordinate. In living systems, time has consequence.

Early in life, correction is cheap. Error is rapidly absorbed. Regulation operates with margin. The system spends little effort maintaining coherence. Time passes lightly because little is at stake.

As regulation loses precision, this changes. Correction becomes more costly. Deviations take longer to resolve. The system spends increasing effort maintaining balance. Time acquires weight.

This weight is not evenly distributed. Periods of stress, illness, or uncertainty compress experience. Attention sharpens. Prediction becomes difficult. The future feels closer. This is not illusion. It is the subjective correlate of operating near regulatory limits.

Time feels slow when uncertainty is high.

This relationship runs deep. Prediction depends on continuity. When prediction becomes unreliable, each moment demands more processing. The system cannot safely extrapolate. Experience fragments.

The same logic explains why routine compresses time. Familiarity stabilises prediction. Correction becomes trivial. Experience smooths.

Ageing alters this balance gradually. Accumulated error degrades prediction. Correction remains possible, but it requires more effort. The system spends more time estimating and less time assuming. This shift is subtle, but persistent.

Time begins to feel faster not because less happens, but because more effort is spent maintaining coherence rather than forming new memory. Experience passes without being deeply encoded. What is not encoded cannot be recalled. What is not recalled seems not to have occurred.

Thus, time accelerates.

This acceleration is not subjective whim. It reflects a redistribution of resources. Memory, attention, and regulation compete. As correction grows more costly, fewer resources remain for novelty.

This also explains why crises imprint so strongly. When systems approach thresholds, prediction fails dramatically. Everything becomes salient. Memory deepens. Time slows.

Noise and accumulation intensify this effect. Variability widens. Rare events become more frequent. The system cannot rely on averages. Each moment demands assessment. Time becomes dense.

Importantly, this density is not pathological. It is adaptive. Near limits, sensitivity is valuable. Systems that fail to slow down near thresholds do not survive.

Time experienced therefore reveals where a system is operating in its mathematical space. Ease corresponds to margin. Strain corresponds to proximity to boundary.

The next chapter will formalise this intuition. Accumulation transforms the landscape in which regulation operates. Drift becomes inevitability. Variance widens. What once felt distant draws near.

Time does not change.
Its cost does.

Chapter 5

Noise and Accumulation

When error becomes inevitable

Regulation corrects deviation, but it cannot eliminate noise. Every living system operates in an environment of continual fluctuation. Signals are corrupted. Measurements are imperfect. Responses are delayed. Noise is not an external insult imposed on life. It is intrinsic to the operation of complex systems.

Noise alone is not catastrophic. What matters is accumulation.

In a regulated system, small deviations are corrected locally and rapidly. Most never propagate far. However, no correction is exact, and no regulatory loop operates in isolation. Each adjustment leaves a residue, however small. Over time, these residues accumulate.

Accumulation does not proceed linearly. Errors interact. Corrections overlap. Delays compound misalignment. What begins as random fluctuation can acquire apparent structure, as corrections overlap and delays align. Noise remains random in origin, but not in consequence.

Mathematically, this widening is unavoidable. A system that corrects error probabilistically will, over repeated cycles, explore an increasingly large region of its possible states. Even when regulation remains effective on average, variance grows. Mean behaviour may remain stable while extremes become more frequent.

Noise accumulates differently across scales. Fast processes generate many small errors that are rapidly corrected. Slow processes generate fewer errors, but corrections arrive late and propagate widely. Interactions between timescales introduce additional instability. What remains robust locally may become fragile globally.

Coordination across scales is therefore a central vulnerability. Living systems depend on timing as much as on magnitude. When delays lengthen and thresholds blur, signals arrive out of phase. Regulation fragments. Components continue to function, but coherence erodes.

The system drifts because it cannot correct itself perfectly forever. Accumulation is informational before it is material. What degrades is not substance, but alignment.

Repair mechanisms slow this process, but they cannot reverse it. Repair restores components, not history. Each repair cycle reduces local error while allowing global variance to grow.

Redundancy, though essential, also has limits. Redundant systems distribute risk, but they increase complexity. More components mean more interactions, more delays, and more opportunities for misalignment. Robustness at short timescales can coexist with fragility at long ones.

Regulation compensates continuously, but compensation itself introduces new error. Control does not fail. It becomes increasingly burdened.

Accumulation therefore sets a horizon. Regulation can delay loss of coherence, but it cannot prevent it indefinitely. The mathematics is uncompromising. Over sufficient time, variance grows and margins shrink.

This horizon is not fixed. It depends on the strength of regulation, the level of noise, and the coupling between scales. Some systems drift slowly. Others reach critical thresholds earlier. The result is heterogeneity rather than uniform decline.

Crucially, accumulation prepares systems for regime change. As variance widens and margins narrow, behaviour becomes increasingly sensitive to perturbation. Small changes that were once absorbed now have disproportionate effects. The landscape has changed, even if the rules have not.

At this stage, the system is not yet unstable. Control remains. Feedback still operates. But the system now lives close to boundaries it once rarely approached. This proximity is decisive.

So far, regulation has held; what changes next is not the presence of noise, but the direction of feedback. In the next chapter, those boundaries will be crossed. When regulation no longer counteracts deviation, feedback reverses sign. Growth accelerates. Instability emerges.

Interpretation

Noise is not the enemy of life. Accumulation is.

Living systems endure not because they eliminate error, but because they postpone its consequences. Over time, postponement becomes insufficient. Variance widens. Margins shrink.

Failure does not arrive when regulation stops working, but when it works too slowly, too late, or with too little authority.

What time does is not to destroy systems outright, but to make rare events common. That shift, once it occurs, cannot be undone.

Equation 5

Diffusion

Why Variance becomes inevitable

In mathematics and physics, this behaviour is known as diffusion, first formalised in Einstein's description of Brownian motion and later reused across biology, finance, and population dynamics.

This equation shows why random fluctuations cause variability to grow over time, even when the average stays the same.

$$\text{Var}(x_t) = \mathbb{E}\big[(x_t - \mathbb{E}[x_t])^2\big] \propto t$$

In ageing, this means that extreme outcomes become more likely long before average function visibly declines.

When small random errors are carried forward over time, the spread of possible outcomes grows, even if the average remains stable. This is diffusion.

This is not universal for all regulated systems. Strong restoring feedback can bound variance. Diffusive widening arises specifically in degrees of freedom that are not fully mean-reverted, or when residual error behaves like integrated noise.

The consequence is subtle but decisive. A system may appear stable by every average measure, yet become progressively more vulnerable as its range of possible states expands.

This is why error accumulation eventually matters.

Here, the collective effect of carried-forward errors becomes visible: the distribution widens.

This widening has profound consequences.

As time passes, the tails of the distribution thicken. States that were once improbable become accessible. Regulation must now contend not only with typical deviation, but with extremes.

Michelangelo's Sistine Chapel ceiling offers a visual analogue of this problem.

As the narrative unfolds across the ceiling, figures enlarge. Early scenes of creation are populated by relatively small bodies set within expansive architectural space. As the sequence progresses, anatomy intensifies. Limbs thicken. Torsos dominate. By the time one reaches the prophets, sibyls, and ignudi, individual figures command entire sections of the visual field.

Local elements grow powerful. Variability increases. The risk of domination by parts is real.

Yet the ceiling does not collapse.

Despite dramatic changes in scale, the composition remains coherent. Proportion is preserved across panels. Rhythms repeat. Balance holds. Enlargement is permitted, but not without constraint. The system tolerates escalation because it is regulated.

This is not expressive excess. It is controlled amplification.

Michelangelo allows figures to grow in prominence only within strict architectural and proportional limits. Scale shifts, but structure remains invariant. Identity persists across transformation. What changes is magnitude, not organisation.

This is the same problem living systems face over time. As variance widens, local processes gain influence. Components become louder. Signals amplify. Without regulation, these changes would fracture coherence. With regulation, the system absorbs them, for a time.

Art achieves this through design. Life must achieve it through feedback.

The ceiling shows what happens when escalation is permitted without surrendering constraint. Biology shows what happens when that constraint weakens. Ageing narrows the margin within which amplification can be tolerated. Cancer marks the point at which it cannot.

What the ceiling makes visible in space, life must negotiate in time.

Diffusion therefore reframes how decline should be understood. The problem is not that the system is performing worse on average. The problem is that **variance has grown beyond what regulation can safely absorb**.

Collapse is triggered not by typical behaviour, but by excursions into the tails.

This is why ageing-related events often feel sudden. The system crosses no visible threshold. It simply encounters a rare but now probable fluctuation.

This equation also explains why interventions become less effective over time. As variance increases, control must work harder to stabilise a wider range of states. Eventually, correction becomes insufficient for the extremes, even if it remains adequate for the average.

Diffusion is impartial. It does not discriminate between systems that are well designed and those that are merely long-lived. It acts wherever noise accumulates over time.

Life persists by narrowing variance through regulation. Ageing widens it inexorably.

Both are true simultaneously.

Consequence

Ageing is not defined by decline in average performance, but by the growth of variability. Risk increases not because systems fail more often, but because rare failures become unavoidable.

This is the point at which drift becomes destiny. What diffusion describes mathematically, history sometimes reveals empirically.

Interlude 7: Change Without Design

The *HMS Beagle* was built to measure coastlines. Its purpose was precision. Depths were sounded, angles fixed, land rendered stable on paper. When it left England in the winter of 1831, it carried instruments designed to reduce uncertainty. It carried no theory of life.

Among those on board was a young naturalist, twenty-two years old, trained to observe rather than explain. His notebooks filled with particulars: shells gathered, bones uncovered, animals compared across distances small enough to cross in hours yet large enough to separate populations completely. He recorded difference without inference.

Patterns did not announce themselves. They accumulated.

In South America, fossil remains appeared in rock, immense in scale yet familiar in form. They resembled living species found nearby, not those elsewhere. Structure persisted. Size did not. Replacement had occurred without redesign.

The land itself shifted. Earthquakes lifted coastlines. Shells appeared above the sea. Change did not arrive as rupture but as repetition. Small movements compounded until they could no longer be ignored. What had seemed permanent proved temporary when viewed across sufficient time.

Time entered the system as an active force.

On isolated islands later in the voyage, variation narrowed. Geography imposed limits. Similar organisms diverged slightly but consistently. Beaks altered shape without abandoning function. Shells curved differently without losing form. The differences were constrained. They repeated.

Nothing guided the process. Nothing needed to.

Organisms poorly matched to local conditions disappeared. Others remained. Nothing was added. What failed was removed.

This was not creation. It was filtration.

Darwin did not possess the language of probability or optimization, but the structure was already complete. Variation introduced difference. Constraint reduced possibility. Time allowed small asymmetries to accumulate. No intention was required. Elimination alone was sufficient.

Much later, the process would be written as an equation:

$$\Delta p \propto \text{variation} \times \text{selection} \times \text{time}$$

Read plainly, it states only that change requires difference, filtered by survival, amplified by duration. It promises nothing else. No progress. No improvement.

When the *Beagle* returned to England, Darwin did not publish. He sorted and compared. He delayed. To accept what he had seen was to accept that persistence required no purpose, and that structure could arise without design.

The same logic governs ageing bodies and dividing cells. It governs populations and tumours adapting within the constraints of their own environments. Selection continues even when it undermines the system that contains it.

Darwin did not impose evolution on life. He encountered it as consequence, time acting on difference under constraint. Mathematics did not revise this insight. It explained why, once such a process begins, there is no mechanism to stop it.

Change does not require intent. It requires only time.

Interlude 8: Why Collapse feels sudden

Collapse is rarely sudden. It is only perceived that way.

Systems that fail abruptly have usually been changing slowly for a long time. Error accumulates. Margins narrow. Variability widens. Yet outward behaviour remains stable. Regulation continues to succeed, until it does not.

This delay between cause and consequence is not deception. It is a property of regulated systems.

As long as deviation remains within correctable bounds, behaviour appears unchanged. Feedback masks underlying deterioration. Performance remains adequate. Signals are compensated. From the outside, nothing seems wrong.

From the inside, the system works harder.

This increasing effort is often invisible. Correction absorbs error before it can propagate. Only when margins are exhausted does behaviour change qualitatively. At that point, small additional perturbations produce large effects. Collapse appears abrupt because the system has crossed a boundary.

Mathematically, this is expected. Systems governed by feedback exhibit nonlinear response near limits. Far from boundaries, perturbations are damped. Near boundaries, they are amplified. The transition between these regimes is sharp.

This explains why warning signs are often missed. Early indicators are corrected away. By the time failure is observable, correction is no longer possible. Collapse is not sudden. Visibility is.

The same logic explains why interventions fail late. When a system has crossed into an unstable regime, restoring prior parameters does not restore prior behaviour. The system has moved into a different region of state space. Reversal is no longer symmetric.

This asymmetry is crucial. Recovery requires more than undoing damage. It requires re-establishing constraint.

This is why collapse feels like betrayal. Long periods of apparent stability create confidence. The sudden loss of coherence seems disproportionate to preceding change. But confidence was an artefact of regulation, not evidence of health.

This dynamic is familiar in many domains. Ecosystems appear stable until populations crash. Financial systems absorb stress until liquidity vanishes. Physiological systems compensate until failure occurs. The form is the same.

What unites these cases is not fragility, but delayed visibility.

Collapse also feels sudden because memory is biased. Gradual change rarely imprints strongly. Only discontinuity is remembered. The long approach to the boundary disappears from narrative.

This bias reinforces the illusion of abruptness. The system did not fail without warning. The warning was not registered.

Understanding this changes how collapse is interpreted. It is not a mystery or a moral failure. It is the predictable outcome of accumulation in regulated systems.

The next chapter will examine what happens after the boundary is crossed. Once correction gives way to amplification, behaviour accelerates. The system enters a new regime.

Instability does not arrive unannounced.
It arrives unnoticed.

Chapter 6

Instability without control

Cancer as a regime change

This chapter does not explain cancer by listing causes, but by describing the moment control changes character.

Ageing prepares the ground. It does not dictate the outcome. Until that point, feedback remains predominantly stabilising and behaviour stays bounded.
Cancer begins when this condition no longer holds.

Understanding cancer as loss of constraint does not make it less devastating; it makes it intelligible.

On imaging, the tumour still responds to oxygen, glucose, and growth signals, yet it expands as if the surrounding tissue no longer exists.

What distinguishes cancer is not growth itself, but loss of constraint. Regulatory systems that once limited proliferation, enforced differentiation, coordinated repair, and integrated cellular behaviour into tissue context cease to function as stabilisers. Feedback reverses sign. Correction becomes amplification.

In a regulated system, negative feedback counteracts deviation. Increasing error strengthens corrective force. Stability is preserved. In cancer, this relationship collapses. Signals that once inhibited growth are ignored, distorted, or actively repurposed.

Pathways that once depended on context acquire autonomy. Positive feedback begins to dominate.

This shift does not require the invention of new biological functions. It arises from reconfiguration of existing ones. The same molecular machinery remains present. What changes is how it is coupled, constrained, and interpreted. Regulation dissolves while activity persists.

Cancer is often described through a list of hallmarks. Mathematically, these describe a single event: the failure of regulatory constraint across multiple dimensions of cellular behaviour.

Correction no longer stabilises deviation; it amplifies it.

Genomic instability, epigenetic plasticity, altered metabolism, inflammation, and immune evasion do not define this transition. They accelerate it by enlarging the space of accessible states once control has weakened.

As constraint erodes, increased variability allows unstable configurations to be explored more rapidly. Identity becomes plastic, metabolism uncouples from organismal need, and immune oversight loses authority. Signals that once confined cells to appropriate spatial and functional niches no longer bind.

Even hallmarks that appear contradictory coexist naturally within this framework. Proliferation and dormancy, expansion and quiescence, plasticity and fixation are not opposing explanations. All represent escape from externally imposed control. Instability does not prescribe a single behaviour. It removes constraint, allowing many.

As cancer is examined across tissues, developmental contexts, and evolutionary timescales, new hallmarks continue to emerge. This expansion does not complicate the picture. It simplifies it. What is expanding is not the number of causes, but the number of dimensions along which regulation can fail.

Tumour suppressors and oncogenes are therefore best understood as complementary components of the same control architecture. Tumour suppressors embody constraint. Oncogenes embody amplification. Cancer emerges when amplification proceeds without counterbalance.

Importantly, instability does not require complete loss of regulation. Partial failure is sufficient. Once net feedback becomes positive, growth accelerates even in the presence of remaining controls. Attempts at correction arrive too late or are overwhelmed. Behaviour escapes the region of bounded response.

This also explains why cancer is difficult to treat. Therapies that suppress activity without restoring constraint do not re-establish stability. The system adapts and persists because amplification continues in the absence of negative feedback.

Ageing widens variance. Cancer collapses boundaries.

In mathematical terms, cancer represents a bifurcation in system behaviour. The governing equations do not change; the system moves into a different solution space.

This regime change explains both the adaptability and the destructiveness of cancer. Unconstrained systems explore state space rapidly and survive perturbation, but they do so at the expense of coherence at the level of the organism.

Interpretation

Cancer is not best understood as excess, but as escape. What is lost is not activity, but constraint.

Systems do not become unstable because they grow too much, but because growth is no longer regulated. When correction becomes amplification, behaviour changes character.

The implication is stark. Stability depends not on suppressing motion, but on preserving control. When control fails, mathematics does not disappear. It becomes destructive.

This framework does not replace molecular oncology; it explains why molecular detail alone cannot restore control once a regime change has occurred.

Equation 6

Positive Feedback

When control reverses

This equation shows what happens when correction is replaced by self-reinforcing amplification. This is the simplest form of **exponential growth**, the same mathematical law that governs unchecked population expansion, compound interest, and nuclear chain reactions.

$$\frac{dx}{dt} = kx, \qquad k > 0$$

In cancer, this is why growth accelerates rather than settling back to a stable baseline, the biological equivalent of a process that no longer slows when it should.

Growth is exponential: the larger x becomes, the faster it increases. This equation looks deceptively similar to the equation of regulation. The sign has changed. The consequences are profound.

Deviation no longer triggers correction. It triggers amplification. Error is no longer opposed. It is rewarded.

This is **positive feedback**.

Positive feedback is not inherently pathological. It is essential for transitions. Development, wound healing, immune activation, and reproduction all depend on temporary amplification. But these processes are safe only because they are embedded within larger regulatory frameworks that eventually reassert control.

Cancer begins when amplification escapes that framework.

In a healthy system, growth is conditional. Signals are integrated. Thresholds are enforced. Negative feedback dominates long-term behaviour. In cancer, this balance reverses locally. Growth signals amplify themselves. Constraints are ignored or dismantled.

This equation captures the essence of malignancy without invoking any specific gene, mutation, or pathway. Those details matter biologically, but mathematically they are mechanisms by which this sign change is realised.

Tumour suppressors, oncogenes, genomic instability, altered metabolism, and immune evasion are not separate explanations. They are different ways of weakening correction and strengthening amplification. They reshape the parameters of this equation.

Once positive feedback dominates, behaviour accelerates. Small differences in initial conditions produce large differences in outcome. Prediction becomes difficult. Intervention becomes reactive rather than preventative.

This explains a central feature of cancer: **nonlinearity**.

Early changes may appear innocuous. Late changes escalate rapidly. Growth curves steepen. Resistance emerges. The system adapts aggressively because amplification favours exploration of state space.

This also explains why cancer feels alien. Its behaviour is no longer coordinated with organismal goals. Yet it is not foreign. It follows mathematics the body already uses, stripped of restraint.

Cancer is not the introduction of new behaviour.
It is the **misapplication of a familiar rule**.

Positive feedback also explains resilience. Systems governed by amplification resist suppression. Attempts to reduce growth are countered by further adaptation. The system does not return to baseline because baseline is no longer stable.

This is why cancer cannot be understood as excessive growth alone. It is **growth without a stable reference point**.

Once amplification replaces correction, the system has crossed a boundary. Regulation no longer defines behaviour. History no longer constrains future trajectories. Growth becomes self-justifying.

This equation therefore marks a qualitative change. It is not an extension of ageing. It is a different regime.

Ageing widens variance.
Cancer escapes constraint.

Consequence

Cancer is not life turned malicious. It is regulation turned unstable.

The same mathematics that once sustained coherence now accelerates divergence. Control has not vanished. It has inverted.

Understanding cancer at this level shifts attention away from isolated causes and toward system-level constraint.

Chapter 7

Phase Transitions in Living Systems

When behaviour changes regime

Up to this point, change has been described in terms of drift, accumulation, and loss of control. These descriptions capture how systems move. They do not yet explain why behaviour sometimes changes suddenly.

An ensemble can tolerate small timing errors until a single bar collapses the rhythm entirely.

The correct mathematical language for this change is **phase transition**.

A phase transition occurs when a system governed by continuous rules begins to behave discontinuously. Small changes in underlying parameters produce large changes in outcome. What changes is the regime in which the system operates.

Living systems exhibit phase transitions not because they are exceptional, but because they are regulated. Regulation allows stability across a range of conditions. It also creates thresholds. When those thresholds are crossed, behaviour changes qualitatively.

Ageing and cancer occupy different sides of this distinction.

This distinction resolves a long-standing confusion: ageing and cancer share molecular features, but differ fundamentally in dynamics, one drifts within control, the other escapes it.

In mathematical systems, phase transitions are characterised by loss of smoothness. Near a critical point, behaviour becomes highly sensitive to perturbation. Small fluctuations that were once absorbed now produce large effects. Prediction becomes difficult. Variability increases sharply.

This behaviour is familiar in physical systems. Water remains liquid across a range of temperatures, then abruptly becomes ice. Magnetisation persists until it vanishes. The underlying interaction laws do not change. What changes is the collective organisation they support.

Living systems behave in the same way. As regulatory precision declines and variance accumulates, systems approach critical thresholds where behaviour becomes increasingly sensitive. Ageing moves systems toward these boundaries without determining which will be crossed. That outcome depends on local conditions, stochastic events, and tissue-specific regulation. Cancer is one possible transition; degeneration, failure, or death are others.

This sensitivity explains why the timing of intervention matters as much as its magnitude. Far from a threshold, even substantial interventions may have little effect because stabilising feedback still dominates. Near a threshold, small changes can redirect trajectories or precipitate collapse.

Seen this way, disease is not simply malfunction, but regime change. Health is not a fixed state, but residence within a stable region of parameter space. The same cellular processes operate in health and disease; what differs is how tightly they are constrained.

Phase transitions are therefore neither mysterious nor exceptional. They are the natural consequence of regulation under pressure.

Interpretation

Living systems do not fail smoothly. They change regime.

Ageing moves systems toward critical thresholds; cancer marks the crossing of one. What appears sudden is the delayed expression of long-term drift finally exhausting regulation.

To understand health, disease, and intervention, it is not enough to catalogue processes. One must know where a system resides in its space of possible behaviours, and how close it lies to a boundary beyond which control cannot be restored.

Stability is never permanent. It is conditional. When conditions change, the nature of life changes with them.

These transitions are not confined to biology. They are already familiar, long before they are named.

A familiar regime change

Love has always embarrassed logic. And yet, whenever thinkers have tried to describe it honestly, they have reached for structure.

In Athens, at a long table heavy with wine and argument, Plato lets the conversation drift toward love. The poets speak first, praising madness and divine possession. Only later do the philosophers respond. Love, Plato suggests, is not a fall but an ascent. It begins with a body, rises toward beauty, and then toward ideas themselves. What appears emotional is, in fact, directional. Love moves. It climbs. It changes form.

The Greeks believed that truth revealed itself through proportion and order. Love belonged to this family. It was not random desire, but a force that followed a path-bending, accelerating, never quite straight. Plato never wrote an equation, but he sketched a curve: desire tending toward the infinite.

Centuries later, Baruch Spinoza attempted something more radical. He wrote about love the way Euclid wrote about triangles. Definitions first. Then axioms. Then propositions that followed with unyielding calm. Love, he said, is joy accompanied by the idea of an external cause. Nothing more. Nothing less.

Many found this unbearable. They wanted love to be mysterious, not necessary. But Spinoza insisted that emotions obey laws whether we acknowledge them or not. We do not choose whom we love freely; we arrive there pushed by prior conditions, memory, temperament, circumstance, timing. When love wounds us, it is not because it is chaotic, but because the system that produced it was misaligned.

Spinoza was expelled from his community for thinking this way. His heresy was simple: he believed love had structure.

In the early nineteenth century, Stendhal noticed something neither Plato nor Spinoza had fully captured. Love does not grow smoothly. It forms suddenly, like crystals along a branch lowered into a salt mine. One glance, one absence, one imagined perfection, and the beloved becomes unavoidable. Reality reorganises itself around them.

Stendhal called this crystallisation. He understood that love does not increase gradually. It waits. Then it tips. Nothing much happens, until everything does.

Much later, scientists would give this behaviour a name. Phase transition. Nonlinearity. Stendhal had only language, but he had accuracy.

By the end of the twentieth century, love entered the laboratory. A psychologist named John Gottman invited couples to sit in rooms wired with sensors. He measured heart rate, facial expression, tone, timing. Within minutes, he could predict which marriages would survive and which would collapse.

Not because of romance. Because of ratios.

Too much criticism without repair, and the system drifted toward instability. Too much contempt, and recovery became impossible. Collapse did not arrive gradually. It arrived suddenly, once a critical threshold was crossed.

Gottman later acknowledged that his models borrowed directly from mathematics. Relationships behave like dynamical systems. Feedback matters. Small signals compound. Stability is fragile. Love does not fade, it flips.

This casts a familiar human experience in a precise light. The person you almost loved but didn't. The one you met too early. The one you met too late. Same individuals. Different outcomes. The difference was not character. It was initial conditions.

What lovers call fate, mathematicians call sensitivity.

Both describe the same unease: small causes, enormous consequences.

Across centuries, love has revealed the same paradox. It is lawful but unpredictable. Structured but uncontrollable. It obeys patterns we feel long before we can name them.

Love is not irrational.

It is non-linear.

Equation 7

Bifurcation

Why behaviour changes regime

This equation shows how a system governed by fixed rules can suddenly change behaviour when a control parameter drifts.

This equation is known as the logistic map, introduced in the study of population growth and later recognised as a universal model of how stability gives way to instability in nonlinear systems.

$$x_{t+1} = r\, x_t(1 - x_t), \;\; x_t \in [0,1], \; r \in [0,4]$$

In ageing and cancer, this explains why long periods of apparent stability can end abruptly once regulatory margin crosses a threshold.
This deterministic nonlinear update rule shows how small changes in a single parameter produce qualitative regime shifts: fixed points, oscillations, period-doubling, and chaos. The rule itself does not change; the behaviour changes because the control parameter r crosses critical values. It is a mathematical demonstration of how gradual parameter drift can yield abrupt changes in outcome.

It is not important because of its specific form. It is important because it shows how **qualitative change** can arise from smooth, continuous rules.

As the parameter r increases, the system behaves predictably at first. It settles into a stable state. Small perturbations are absorbed. Behaviour is regular.

Then something changes.

At a critical value of r, stability is lost. The system begins to oscillate between two states. Increase r further, and the oscillations double. Then double again. Eventually, behaviour becomes chaotic. Prediction fails. Small differences in initial conditions lead to entirely different outcomes.
Only the parameter has shifted.

This is a **bifurcation**.

Bifurcations explain why systems governed by continuous laws can behave discontinuously. They show why gradual change in underlying conditions can produce sudden, dramatic shifts in behaviour.

For long periods, behaviour remains stable. Then, abruptly, it does not.

Ageing moves systems along the parameter axis.
Cancer marks the crossing of a critical point.

Ageing changes parameters. Cancer changes the regime itself.

The logistic map also reveals asymmetry. Once a system has crossed a bifurcation, returning parameters to previous values does not necessarily restore previous behaviour. History matters. Paths are not reversible. The system remembers.

This irreversibility is crucial. It explains why late intervention often fails. The system has reorganised itself. Constraint has been lost not incrementally, but structurally.

Bifurcations also explain sensitivity. Near critical points, small perturbations have outsized effects. Variability explodes. Prediction becomes unreliable. Control strategies that once worked now destabilise the system further.

This behaviour is not pathological. It is intrinsic to nonlinear systems.

The logistic map teaches a sobering lesson. Stability is not guaranteed by good design. It is conditional on parameters remaining within narrow bounds. Cross those bounds, and behaviour transforms.

This is not failure of regulation.
It is the limit of what regulation can achieve.

Consequence

Systems do not fail gradually. They change regime.

Ageing erodes margin. Accumulation widens variance. Positive feedback accelerates divergence. Bifurcation marks the point at which coherence can no longer be restored by incremental correction.

Understanding disease therefore requires more than identifying causes. It requires knowing where a system sits relative to its critical thresholds.

Once crossed, the problem is no longer repair.
It is containment, redirection, or replacement.

Interlude 9: Diffusion beyond biology

The same mathematics that widens variance in ageing operates wherever systems persist under noise for long periods.

What diffusion produces in ageing, widening variance without visible decline, it also produces in institutions that persist across centuries.

Large-scale outcomes are often attributed to large-scale forces. Mathematics suggests the opposite. In complex systems, divergence is driven by small, repeated asymmetries operating at fine resolution. When such asymmetries persist over long time scales, error becomes not a possibility but a certainty. Oxford and Cambridge provide a rare opportunity to observe this process directly, because many of these asymmetries are formalised, stable, and openly visible.

The relevant time horizon is unusually long. From the early thirteenth century onward, English governance shifted from unconstrained authority to constrained authority, formalised in instruments such as Magna Carta. This shift did not flatten the distribution of power; it stabilised it. Authority was bounded and redistributed, but not equalised. In mathematical terms, volatility was reduced while asymmetry was preserved. The system traded fluctuation for persistence.

The collegiate system mirrors this logic at smaller scale. It does not merely partition students administratively; it differentiates their daily environments. Colleges vary in endowment, architecture, proximity to departments, and historical density.

No single difference dominates. Together, they alter the variance of trajectories.

Even routine practices participate in this process. Dining arrangements differ within and between colleges. Faculty dine at High Table; students dine below. This separation is not decorative. It establishes a stable informational gradient. Authority is spatially encoded, reducing uncertainty about where evaluation resides. Students need not infer the location of influence; it is visible.

Interaction across this boundary is infrequent, but frequency is not the controlling parameter. In systems sensitive to early perturbation, sparse signals outweigh later noise. A single formal remark, delivered early and ceremonially, can redirect a trajectory more effectively than weeks of anonymous correction. Control is achieved not through volume, but through timing.

The barrier is not procedural but psychological, and therefore efficient. Access exists nominally for all; activation energy varies silently. Those comfortable approaching asymmetry do so; those who are not, self-select away. No exclusion is required. The filter operates automatically.

Mentorship follows the same pattern. Colleges differ in tutorial intensity, senior availability, and alumni engagement. These differences function as feedback controls. Where feedback is frequent and local, variance narrows. Where it is delayed or diffuse, variance spreads. The effect is cumulative. One conversation alters little; repeated correction reshapes phase space.

Financial structure compounds the divergence. Departments standardise academic content and assessment; colleges govern lived conditions, buffering, and informal networks. This separation preserves systemic stability while allowing heterogeneity of experience. Equality of rules does not equalise trajectories.

Such differentiation shapes more than academic performance. Confidence, institutional fluency, and procedural ease evolve in parallel. Students habituated to formal ritual, asymmetrical interaction, and institutional continuity learn not only subject matter, but how systems respond to pressure. This knowledge generalises.

The result is not destiny, but probability. Graduates from certain colleges are not guaranteed influence, but they are statistically more likely to encounter pathways with lower friction. When combined with dense networks, such as debating societies, alumni structures, and historical ties to governance, these probabilities compound.

Crucially, none of this requires exclusion. Admissions may widen while divergence persists. Equal entry criteria do not equalise outcomes because experience is multidimensional. Mathematics predicts this result. Growth across partitions with unequal resources stretches distributions even under neutral rules.

Over long durations, micro-asymmetries aggregate into macro-patterns. Certain colleges become statistically over-represented in public life, science, and governance. The pattern is visible not because it is engineered, but because it is consistent. Systems that reduce uncertainty early produce trajectories that appear disproportionately successful later.

Biology provides the limiting case. Cells exposed to marginally different signalling environments diverge over time, even when genetically identical. Differentiation is not malfunction; it is structure unfolding. Institutions behave similarly.

What appears as selection is often diffusion observed late. Possibility widens quietly; collapse begins when regulation can no longer contain it.

Chapter 8

Health as Dynamic Balance

Stability without Stillness

After instability, it is tempting to define health as its opposite. This temptation should be resisted. Health is not the absence of disease, nor is it a return to some original state. It is a condition of **dynamic balance**.

Living systems are never static. They fluctuate continuously. Signals vary. Components turn over. Environments change. Health does not require stillness. It requires that these fluctuations remain constrained within ranges that preserve coherence at the level of the organism.

A healthy heart does not beat at a single rate: it accelerates with exertion, slows with rest, and returns without damage.

Mathematically, health is not a point. It is a region.

Within this region, regulation remains effective. Deviations are detected. Corrections arrive in time. Feedback remains predominantly stabilising. Variability exists, but it is bounded. The system moves, but it does not escape.

This framing resolves a common misconception. Health is often equated with optimisation. Systems tuned too precisely to a single state lose flexibility. They respond poorly to unexpected perturbation. Health depends not on maximising performance, but on maintaining margin.

Margin is what allows error without collapse.

Living systems tolerate inefficiency, redundancy, and excess capacity because these preserve margin; what appears wasteful under optimisation is protective under regulation.

This perspective also clarifies why health declines gradually rather than abruptly. As regulatory precision erodes and noise accumulates, the region of stability narrows. The system remains within it for long periods, but with decreasing room to manoeuvre. Perturbations that were once absorbed now produce disproportionate effects.

Health is lost not when fluctuation appears, but when fluctuation exceeds available margin.

Importantly, dynamic balance is context-dependent. What counts as stable varies with environment, developmental stage, and demand. A healthy system adapts its operating range rather than clinging to fixed values. Flexibility is not a threat to health. It is one of its defining features.

This explains why interventions that aim to restore specific targets often disappoint. Returning individual variables to youthful levels does not necessarily restore balance. Health depends on coordination across scales, not on isolated parameters.

Restoring balance therefore requires attention to regulation rather than replacement. Enhancing feedback, reducing unnecessary noise, and preserving margin are more effective than forcing individual components into predefined states.

Health is thus not a static achievement. It is a **continual process**. Regulation must be renewed. Coordination must be maintained. Balance must be actively preserved.

Crucially, balance does not imply symmetry. Living systems often operate asymmetrically, prioritising some functions over others depending on circumstance. Health lies not in uniformity, but in appropriate response.

This reframing dissolves the boundary between health and disease. Both arise from the same mathematics. What differs is the region of operation. Disease is not the presence of new rules. It is residence outside the region where existing rules preserve coherence.

Seen this way, health is not fragile because it is delicate. It is fragile because it is conditional. It persists only so long as regulation remains effective and margin remains available.

The task of medicine, and of living systems more generally, is not to eliminate change, but to keep change within bounds. Health is not preserved by eliminating risk, but by preserving the capacity to absorb it.

Interpretation

Health is not stillness, and it is not perfection. It is the ability to remain coherent while continually changing.

To care for health is therefore not to seek fixed targets, but to preserve balance. Stability is not something to be achieved once. It is something that must be continually renewed.

Equation 8

Boundedness

What health actually is

This equation defines health as remaining within limits rather than holding a fixed value.

$$x(t) \in [x_{min}(t), x_{max}(t)]$$

This idea appears throughout dynamical systems theory and control theory, where stability is defined not by exact values but by remaining within allowable bounds despite fluctuation.

In real life, this means resilience comes from margin, not optimisation.

Health is not a point but a **viable region**: variables can fluctuate widely and still remain functional as long as they stay within bounds. The bounds are **context-dependent**, exertion, environment, illness, and age change the permissible range, so the region itself can widen or narrow over time.

This distinction corrects a deep misunderstanding. Health is often imagined as precision: the right value, the correct number, the ideal state. In reality, precision is fragile. Systems that require exactness fail easily. Living systems survive by tolerating variation.

The equation above makes this explicit. What matters is not where the system is, but whether it remains within its permissible range.

Boundedness is therefore the mathematical definition of health.

Within bounds, regulation remains effective. Deviations are detected and corrected before they propagate. Feedback dominates amplification. Noise is absorbed. Identity persists.

Outside bounds, regulation fails. Correction arrives too late or not at all. Positive feedback may dominate. Instability accelerates. The system changes regime.

This framing dissolves the false opposition between stability and flexibility. A bounded system can be highly dynamic. Heart rate varies. Hormone levels oscillate. Neural activity fluctuates. Health does not suppress these movements. It constrains them.

Importantly, bounds are not fixed. They adapt to context. Physical exertion widens acceptable ranges. Illness narrows them. Ageing gradually compresses margin. Health is therefore conditional, not absolute.

This conditionality explains why identical interventions produce different outcomes. A perturbation that is harmless in a wide-margin system may be catastrophic in a narrow-margin one. The difference lies not in the perturbation, but in available space.

Boundedness explains why inefficiency is protective. Redundancy, reserve capacity, and slack appear wasteful only under optimisation; in bounded systems, they preserve distance from critical thresholds.

This is why perfect optimisation is dangerous. Systems tuned to operate at their limits lack room for error. When disturbance arrives, there is nowhere to go.

Health therefore resists optimisation. It prefers margin.

This equation also clarifies prevention. Preventing disease is not about eliminating every risk. It is about preserving bounds. Systems fail not because perturbations occur, but because there is no longer space to absorb them.

In this sense, health is not a static achievement. It is an **ongoing negotiation** between regulation, noise, and time. Bounds must be actively maintained. They shrink if neglected. They cannot be restored instantly once lost.

The elegance of this equation lies in its humility. It does not promise permanence. It does not define an ideal. It describes what is required for coherence to persist.

Health is not perfection.
It is remaining inside the region where correction is still possible.

Consequence

To care for health is not to chase optimal values, but to protect margin. Stability does not require stillness. It requires room to move without falling apart.Living well is therefore not about control alone, nor about freedom alone. It is about remaining within bounds while continuing to change.

This balance is temporary.
That does not make it meaningless.

Interlude 10: What Mathematics Cannot Save

Mathematics describes structure, constraint, and change with extraordinary precision. It reveals what is possible, what is stable, and what is inevitable. Yet there are limits to what description alone can preserve.

Mathematics can explain why systems drift. It can show how error accumulates and why collapse occurs. It can identify thresholds and predict regime change. What it cannot do is stop time from acting. Mathematics exposes necessity; it does not negotiate with it.

Living systems persist by regulation, not by understanding alone. They correct deviation without knowing why. They adapt without foresight. Meaning arises from continued coherence, not from awareness of its fragility. Understanding often arrives too late to save the system it describes.

This asymmetry is uncomfortable. We are inclined to believe that knowledge confers control. In complex systems, the opposite is frequently true. The more clearly instability is understood, the more evident it becomes that some transitions cannot be reversed once crossed.

This does not make understanding futile. It changes its purpose.

Mathematics cannot restore youth, eliminate uncertainty, or guarantee stability. What it can do is clarify where intervention is possible and where it is not. It can distinguish delay from reversal, and prevention from cure.

This boundary is not unique to biology. It appears within mathematics itself. In the twentieth century, Gödel showed that any sufficiently powerful formal system contains true statements that cannot be proven within the system, and that such a system cannot establish its own consistency using its own rules. Mathematics, pursuing certainty with unmatched discipline, encountered a boundary that further rigor could not cross.

This was not a defeat. It was an honesty.

Long before this result was formalised, many philosophical and religious traditions organised human life around a similar recognition: that ultimate justification lies beyond formal derivation. Law could be fixed while interpretation remained open. Order could be affirmed while essence remained inaccessible. Stability could arise without permanence. The convergence here is structural, not doctrinal. Mathematics did not confirm religion, nor did religion anticipate mathematics in any technical sense. Both encountered the same limit: that formal systems do not justify their own beginnings nor decide their own ends.

Modern medicine inherits mathematics at precisely this point of maturity. Models predict disease trajectories, optimise treatments, and extend life, often dramatically. But they cannot decide when extension ceases to serve the person living it. They cannot weigh duration against dignity, or intervention against suffering. At that boundary, calculation yields to judgment.

This is not a failure of science. It is the moment where science completes its task.

Mathematics tells us what follows from what we assume. It does not tell us which assumptions are worth sustaining. That decision has always belonged elsewhere, to ethics, to culture, to traditions that learned to live without final proofs.

Mathematics can preserve systems.
It cannot decide which ones should be preserved.

The final chapter will gather what has been established and release it. The equation of life does not console. It clarifies. And clarity, when nothing else remains, is still something.

Chapter 9

The Equations of Life

What Persists, What Drifts, What Breaks

The scan has not changed, but the patient has, and "stable" now means something different. This book began with a simple observation: the same patterns recur across music, life, ageing, and cancer. What differs is not the mathematics, but the conditions under which it operates.

It ends with an equation.

Not an equation written in symbols, but one expressed in behaviour.

Life exists where **structure**, **regulation**, **noise**, and **time** interact within limits that preserve coherence. When those limits are respected, systems endure. When they are approached, systems drift. When they are crossed, systems change regime.

Across music, life, ageing, and cancer, the mathematics does not change, only the tolerance for deviation does.
Health is not perfection. It is dynamic balance.

At no point does mathematics disappear. It changes role. It governs rhythm in music, regulation in life, fragility in ageing, and breakdown in cancer.

This perspective dissolves false boundaries. Biology does not stand apart from physics. Disease does not violate natural law. Life does not escape mathematics. It inhabits it.

Living systems endure because they are neither rigid nor random. They operate within narrow corridors where correction is possible and variability is tolerated. These corridors are not permanent. They narrow with time. They can be breached. But while they exist, they sustain coherence.

What we call meaning arises within these same constraints. Expectation, prediction, satisfaction, and loss all depend on structure persisting long enough to be recognised. Music moves us because it balances repetition and surprise. Life matters because it holds together in the presence of uncertainty.

The lesson is not one of pessimism. Drift is inevitable, but collapse is not immediate. Accumulation sets horizons, not deadlines. Instability is catastrophic only when control is lost entirely.

The deeper lesson is one of attention.

To understand life is to listen for its mathematics. To recognise when rhythms are aligned, when feedback is weakening, when noise is accumulating, and when thresholds are near. This applies not only to cells and tissues, but to systems of every scale.

Nothing lasts forever. But coherence can last long enough to matter.

And mathematics, in the end, is not about numbers. It is about what holds, what drifts, and what breaks.

Interpretation

The equations of life do not promise permanence. They describe possibility.

To live well is not to escape these equations, but to remain within their bounds for as long as coherence can be sustained. Balance is temporary. Regulation is costly. Drift is unavoidable.

Life persists not by defying mathematics, but by inhabiting it, briefly, imperfectly, and beautifully.

Equation 9

The Equation of Life

What Endures, What Drifts, What Breaks

A Balance Law

Everything in this book has been converging on a single relationship rather than a final answer.

This equation expresses life as a balance between regulation and disturbance, defining when coherence can persist and when it must decline.

$$\frac{dc}{dt} = R(C, S, t) - n(t), \qquad R \geq 0, \; n(t) \geq 0$$

Equations of this kind appear throughout science under different names. In physics, they describe conservation laws. In engineering, they appear as balance equations governing stability and control. In ecology, they capture the tension between growth and depletion. In neuroscience, they describe excitation held in check by inhibition.

What changes across these fields is the interpretation of the terms. What does not change is the structure of the relationship.

In lived terms, life endures only while regulation can keep pace with accumulated noise; ageing reflects gradual imbalance, and cancer reflects local collapse of control.

This equation is **schematic**: its terms denote relationships rather than directly measurable quantities. It is not a predictive model but a **balance law**. In this sense, it resembles the balance laws that underpin thermodynamics and control theory: not recipes for prediction, but statements of what must hold if coherence is to persist.

Here, $C(t)$ is a coarse measure of coherence, functional organisation that persists over time. $R(C, S, t)$ represents regulatory capacity acting on available structure S, while $n(t)$ represents the instantaneous burden of noise and disturbance.

Coherence increases when regulation outweighs disturbance and decreases when disturbance dominates. Ageing corresponds to a long-term shift in this balance as regulatory precision erodes. Cancer corresponds to a local collapse or inversion of effective control.

This equation does not predict outcomes. It states the **minimal conditions under which coherence can persist at all**.

Coherence increases when regulation successfully acts on structure, and decreases as noise accumulates.

Regulation acts only on what can be preserved; structure defines the space within which correction is meaningful.

Written symbolically, the same relation can be read as:

$$\text{Life} \approx \text{Structure} \times \text{Regulation} - \text{Noise} \times \text{Time}$$

It makes explicit what the formal equation already implies: life persists where structure and regulation can offset the cumulative effects of noise over time.

Ageing corresponds to gradual dominance of the subtraction term.
Cancer corresponds to collapse of the regulatory term.
Health corresponds to maintaining the inequality in favour of coherence.

This is not an equation to be solved.
It is an equation to be **recognised**.

You are not outside this equation, and you are not its subject. You are one of its temporary solutions.

It gathers everything that has come before into a single relation. Nothing here is optional. Nothing can be removed without altering the meaning.

Structure defines what is possible.

Regulation determines what persists.

Noise introduces variation.

Time accumulates consequence.

Life exists where these terms remain in balance.

This equation explains why life is neither static nor chaotic. It persists by correcting deviation while tolerating uncertainty, not by eliminating noise, but by preventing noise from overwhelming structure before time takes its toll.

For this reason, decline is inevitable without being immediate. Noise acts continuously, but its effects are delayed. Regulation compensates. Structure holds. Time passes. Only later does imbalance become visible.

Cancer is what this equation looks like when regulation locally collapses and effective feedback inverts.

Health is what this equation looks like when margins are preserved.

Quantum systems obey structure without regulation. Artificial systems implement regulation without life. Living systems alone bear the full equation. They persist under noise, through time, without guarantee.

This is why life is fragile.
This is why life is resilient.
This is why life matters.

The equation does not promise permanence. It describes conditions under which coherence can exist at all. It tells us why stability is costly, why collapse feels sudden, and why balance is temporary.

Nothing in this equation is moral, sentimental, or consoling. It does not tell us how to live.

It tells us **why living feels the way it does**.

To experience anticipation, effort, fatigue, hope, and loss is to inhabit this equation from the inside. Meaning arises not despite these constraints, but because of them. What can be lost can matter. What must be maintained can be valued.

Life does not escape mathematics.
It unfolds within it.

And mathematics, at its deepest level, is not about numbers or symbols. It is about what can endure long enough to be recognised.

Final Consequence

The equations of life do not guarantee survival. They explain possibility.

To live is to remain, for a time, on the favourable side of this balance, preserving structure, sustaining regulation, absorbing noise, and negotiating with time.

That balance is temporary.
That is not a flaw.

It is the condition under which coherence, meaning, and life itself can exist.

If the mathematics has done its work, you should now recognise these patterns even where no equations are written.

The same balance law applies wherever regulation confronts noise over time.

Interlude 11: Markets fail for the same reason bodies do:

They optimise for the present using assumptions that quietly expire.

The room was calm. Too calm.

It was late summer in **Greenwich, Connecticut**, not the Greenwich that set the world's clocks, but the one that borrowed the name. Like many financial structures, it carried the authority of an older system without having built it.

The screens were on. The phones were quiet. Prices moved, but not violently. The models ran continuously. Every number remained within tolerance.

This was how it always looked before it stopped working.

At Long-Term Capital Management, risk did not arrive as alarm. It appeared as small deviations, fractions of a basis point, movements too small to matter alone. Most days, those differences vanished by afternoon. The equations said they would. They always had.

Mean reversion was not a belief. It was an observed regularity, measured across markets and decades.

Some of the people watching those screens would later receive the Nobel Prize in Economics. Their work defined modern theories of risk and pricing.

Markets were open. Governments functioned. Time, calibrated elsewhere, by another Greenwich, moved forward.

Inside the system, something had shifted.

Russia had defaulted days earlier. The news passed across the screens, registered, and was classified as external. The firm's positions were global, diversified, hedged. Local failures were noise. That assumption had been tested before.

What had not been tested was simultaneity.

By midweek, trades that had never moved together began to align. Bonds meant to cancel one another's risk declined in unison. Liquidity thinned without disappearing. Orders still executed. Prices still printed. But exits narrowed.

No equation failed. What changed was correlation.

Leverage, invisible during years of stability, became suddenly present. Small price movements translated into real losses. Positions were reduced where possible, but selling pushed prices further against them. Each adjustment tightened the next.

The models updated. The numbers remained plausible. Nothing instructed them to stop.

By the time human judgment overrode mathematics, the system no longer had room to respond. Margin calls arrived not as shocks, but as confirmations. The space between what the models allowed and what the market demanded had closed.

Across the system, similar calculations were being made. Hedges behaved identically. Diversification revealed itself as symmetry. Risk had not disappeared. It had aligned.

Weeks later, the Federal Reserve convened not to debate theory, but to contain propagation. The concern was not a hedge fund. It was coupling. The system had become too efficient to absorb failure locally.

Nothing about the collapse was mysterious. The mathematics was sound. The data was accurate. The optimisation had succeeded, until the conditions that justified it expired.

Biological systems fail this way. Cells optimise under stable conditions. Repair mechanisms compensate for common damage while rare errors accumulate unnoticed. Stability persists until correlation overwhelms control, and failure feels sudden only because its preparation was silent.

Markets are living systems in this sense. They do not fail because they ignore mathematics. They fail because mathematics, applied too successfully, removes the slack that once made error survivable.

Greenwich, Connecticut did not look like a disaster zone. It looked like competence.

It always does.

Epilogue

Beyond Life

Quantum systems and Artificial Intelligence

Anyone who has watched a perfectly specified algorithm repeat the same mistake without learning has already seen a system governed by exact rules without memory or correction.

The equation developed in this book does not end with biology. It describes how systems behave whenever structure, regulation, noise, and time interact. Seen from this perspective, both quantum systems and artificial intelligence offer illuminating contrasts.

Quantum mechanics reveals what happens when structure exists without classical regulation, before environmental coupling imposes effective control. At the quantum scale, behaviour is governed by precise mathematical laws, yet outcomes are intrinsically probabilistic. Particles do not follow trajectories. They occupy distributions. Measurement collapses possibility into outcome. Quantum mechanics gives us the Schrödinger equation: exact, deterministic in form, yet yielding probabilities rather than trajectories.

This is not randomness in the everyday sense. It is constraint without determinism.

Quantum systems are exquisitely structured. Their equations are exact. Yet they lack the stabilising feedback that characterises living systems. There is no correction of deviation over time. No accumulation of history. No memory. Quantum behaviour does not drift. It does not age. It simply evolves according to rule until observation intervenes.

Life, by contrast, exists because it remembers. Regulation depends on history. Error accumulates. Correction is informed by past deviation. Time matters.

Stephen Hawking: Structure Without the Body

Stephen Hawking offers a revealing limiting case.

Trained at Oxford but intellectually formed at Cambridge, Hawking gravitated toward a mathematical culture that valued structure over supervision and formal constraint over interpretive dialogue. This preference is sometimes described institutionally, as a matter of academic style. In truth, it became existential. As his disease progressed, the ordinary regulatory capacities of the body collapsed almost entirely. Speech, movement, and autonomous correction disappeared. What remained intact was mathematical structure.

Hawking's work did not depend on bodily regulation in the way experimental practice does. It depended on coherence sustained over time: equations that remained valid regardless of the condition of the organism articulating them. As biological control failed, external systems substituted for it. Assistive technology replaced speech. Collaborators carried calculation. Institutions provided continuity. Regulation did not vanish; it was displaced outward.

Life persisted only through massive scaffolding. Mathematics persisted without it.

This asymmetry is instructive. Mathematics can endure where biological regulation cannot. It can remain exact when the body is no longer capable of correction, memory, or repair. Life, by contrast, cannot. Hawking's survival required continual external regulation, because living systems must preserve coherence under noise and time. When internal regulation fails, persistence depends on whether control can be supplied from outside.

His career demonstrates both the power and the limit of mathematics. Mathematical structure does not age. It does not drift. It does not accumulate error. But life does. The price of inhabiting mathematics rather than merely describing it is vulnerability.

Hawking's work reminds us that mathematics is not life, even when it outlasts the body that holds it. Structure can survive collapse. Regulation cannot always follow. Life exists only while the two remain coupled.

Artificial Intelligence: Regulation without Life

Artificial intelligence presents the opposite case. Here, regulation exists without life.

Modern AI systems operate through layered feedback, error correction, and optimisation. They learn by minimising deviation between prediction and outcome. In this respect, they resemble regulated systems. Yet they lack intrinsic constraint. Their goals are imposed externally. Their regulation is instrumental, not self-preserving.

When AI systems fail, the cost is externalised, to users, institutions, or societies, because the system itself has nothing at stake.

AI systems do not age in the biological sense. They degrade only when hardware fails or objectives change. Noise does not accumulate unless it is allowed to. There is no endogenous drift unless it is engineered. Regulation does not protect identity. It serves performance.

This distinction is crucial. Living systems regulate in order to persist. AI systems regulate in order to optimise. One preserves coherence. The other improves output.

When AI systems fail, they do not drift. They misgeneralise, overfit, or amplify error according to objective functions that remain intact. Instability in AI is not a regime change of the system itself, but a mismatch between optimisation and the context in which it is deployed.

Quantum systems show us structure without correction. AI shows us correction without life. Living systems occupy the narrow region where both coexist.

Life as Regulated Persistence

This comparison sharpens the book's central claim. Life is not defined by computation, intelligence, or complexity. It is defined by **regulated persistence under noise and time**.

Quantum mechanics shows that mathematics can describe the deepest structure of matter without producing life.
Artificial intelligence shows that regulation can generate complex behaviour without producing life.
Life emerges only where regulation acts on structure in the presence of unavoidable noise over extended time.

This is why ageing exists. This is why cancer exists. This is why health is fragile.

It is also why life is rare.

Final reflection

The equation of life does not belong exclusively to biology. It describes a class of systems that must endure rather than compute, persist rather than optimise, and remain coherent rather than exact.

Quantum systems obey mathematics without memory.
Artificial systems optimise without vulnerability.
Living systems do both, and pay the price.

Life exists in the narrow space where error cannot be eliminated, control cannot be perfect, and time cannot be stopped.

That space is small.
That space is costly.
That space is where meaning arises.

On living against fixed odds

The mathematics described in this book can appear unforgiving.
Drift is inevitable. Noise accumulates. Regulation weakens.
Instability emerges. From this perspective, life seems to move
inexorably toward loss.

Yet this conclusion rests on a subtle misunderstanding.

Mathematics does not describe destiny. It describes **possibility
under constraint**. Probabilities shift. Boundaries move.
Thresholds depend on context. The odds are never fixed.

Living systems persist precisely because they operate far from
certainty. Regulation does not eliminate risk. It reshapes it.
Prediction does not guarantee outcome. It biases it. Life
continues not because collapse is impossible, but because it is
continually deferred.

In many physical systems, small changes in conditions can
produce radically different outcomes. Near critical points, tiny
perturbations redirect trajectories entirely. The same
mathematics that makes collapse inevitable under some
circumstances also makes escape possible under others. Even
physics has learned this lesson repeatedly. Boundaries once
thought absolute have softened with deeper understanding. Black
holes, long assumed to be perfect sinks from which nothing could
return, are now known to leak. What appeared final turned out
to be conditional. Living systems exist in this same space. Drift
narrows margin, but it does not abolish possibility.

Living systems are never guaranteed survival, but neither are they condemned by statistics alone. The future is not a straight extrapolation of the past. Possibility persists precisely because conditions, not laws, determine outcome.

To live optimistically, in this sense, is not to deny mathematics. It is to understand it deeply.

Optimism is not the belief that things will improve. It is the recognition that **systems near limits are sensitive**, and that sensitivity cuts both ways. Small actions matter most when margins are thin. Intervention is most powerful near thresholds.

Life persists by inference. It anticipates, adjusts, and acts under uncertainty. It commits resources without certainty of return. It continues despite incomplete information. What we experience as hope is the subjective correlate of this behaviour.

Meaning arises not because outcomes are assured, but because they are not.

To live within these equations is to accept that coherence is temporary, balance is fragile, and control is costly, and yet to act anyway. Not because success is guaranteed, but because the odds are always conditional, and therefore always open to change.

Mathematics does not promise permanence.
It permits possibility.
And that, for life, is enough.

Appendix

Mathematics under constraint: When Time determines outcome

This appendix gathers several historical episodes that do not belong in the main narrative of this book, but which illuminate its central claim under extreme conditions. In each case, mathematics did not merely describe events after the fact. It altered what was possible by shaping how uncertainty, delay, and instability were managed in real time.

These are not metaphors. They are demonstrations.

A. The Atlantic, 1942: Information Decay and Regime Change

The Battle of the Atlantic was not decided solely by ships or weapons, but by whether uncertainty could be reduced quickly enough to matter. German U-boats did not require perfect information; they needed only to raise encounter probabilities beyond what Allied shipping could sustain. The ocean functioned as a probabilistic field, governed by incomplete information and delayed response.

In February 1942, the German Navy introduced a new four-rotor Enigma cipher. This change did not make Allied cryptanalysis impossible. It made it slow. Messages continued to be intercepted and partially decrypted, but no longer within the time window required to influence convoy routing reliably. Knowledge arrived after action rather than before it.

This distinction is mathematical. Information has a half-life determined by the interval in which it can still alter feasible action. Intelligence that arrives late is not intelligence at all.

The blackout that followed was therefore not gradual degradation but a regime change. Regulation failed not because information vanished, but because inference could no longer be completed in time. Shipping losses rose sharply, not because U-boats became more numerous, but because uncertainty could no longer be bounded quickly enough to permit control.

The recovery of cipher material from U-559 in late 1942 restored missing constraints and collapsed the cryptanalytic search space. Within weeks, British codebreakers regained the ability to infer U-boat patrol lines fast enough to act. The Atlantic did not become safe, but it became regulated again. Shipping losses declined not because submarines disappeared, but because uncertainty was once more constrained in time.

This episode illustrates a central theme of this book: **delay and control are not the same achievement**. The Enigma blackout was not a gradual loss of knowledge. It was an abrupt transition caused by the failure of timely inference. When constraints returned, control returned with them.

Mathematics did not sink submarines.
It changed which ships met them, and when.

B. The Manhattan Project: Thresholds are not approached, they are crossed

The Manhattan Project confronted a problem unlike those discussed so far. It was not concerned with gradual accumulation, slow drift, or delayed correction. It faced a hard boundary.

A nuclear chain reaction does not scale smoothly. Below a critical threshold, it dies out. Above that threshold, it grows explosively. There is no stable middle ground.

The central mathematical question was therefore not *whether* energy could be released, but *whether the system could be forced across the critical boundary fast enough* that competing processes, mechanical disassembly, heat expansion, and deformation, could not intervene to stop it.

This was a problem of timing, geometry, and symmetry, not of accumulation.

The failure of the plutonium gun-type design in mid-1944 illustrates this clearly. Plutonium emits spontaneous neutrons. In a slow assembly, these neutrons would initiate the reaction too early, causing the system to blow itself apart before reaching criticality. The result would not be a small explosion, but no explosion at all.

The solution, implosion, was not an incremental improvement. It was a structural reconfiguration. By compressing the plutonium rapidly and symmetrically, density increased faster than disassembly could occur. Reactivity rose abruptly, not gradually.

This introduced a new class of problem. Success now depended on the precise coordination of converging shock fronts. Slight asymmetry would prevent criticality from being reached. The challenge was not engineering alone, but applied mathematics: nonlinear dynamics, hydrodynamic instability, amplification under compression, and robustness under uncertainty.

Design calculations did not aim for an optimal value. They aimed for certainty of threshold crossing. The device was built not to be efficient, but to be *unmistakably supercritical*.

The Trinity test demonstrated a fundamental principle. Approaching a boundary is not the same as crossing it. Below criticality, nothing happens. Beyond it, return is impossible.

What followed at Hiroshima and Nagasaki was not a further calculation, but the consequence of a boundary already crossed. Mathematics had finished its work before the plane ever took off.

The Manhattan Project belongs here because it provides the clearest historical demonstration that **phase transitions are qualitatively different from gradual change**. Systems can drift for long periods without visible consequence. Outcomes transform only when a boundary is crossed.

The same logic governs biological ageing, disease, and cancer. Regulation can weaken gradually. Variance can widen slowly. Yet transformation occurs abruptly, when control fails and a new regime becomes unavoidable.

C. Apollo 11: Regulation without margin

Spaceflight exposes control in its purest form. There is no environment to absorb error, no intuition to rely on, and no tolerance for drift. In orbit, mathematics does not merely describe motion. It governs it.

The Apollo Guidance Computer operated under extraordinary limitations by modern standards: minimal memory, limited processing power, delayed inputs, and noisy sensors. Yet it sustained continuous real-time regulation of a nonlinear dynamical system under uncertainty. It did not compute exact trajectories from first principles. It estimated state, predicted deviation, and corrected continuously.

This distinction matters. Apollo was not guided by calculation alone, but by regulation.

During the Apollo 11 lunar descent, the guidance computer began issuing overload alarms. By any naive metric, the system was failing. Tasks were being dropped. Capacity was exceeded. Yet the system survived not by perfection, but by prioritisation. Nonessential computations were discarded. The guidance loop, the function that preserved bounded descent, was maintained. Control held.

This was not heroism, nor improvisation. It was mathematical discipline embedded in design.

The system was built to preserve what mattered when margin vanished.

The return to Earth exposed the same principle in reverse. Re-entry required navigation through a narrow atmospheric corridor measured in degrees. Too shallow, and the spacecraft would skip back into space. Too steep, and it would burn. Survival depended not on strength or redundancy, but on remaining within bounds after days of accumulated error.

Apollo demonstrates that regulation, not robustness, sustains systems where margin is minimal. The spacecraft did not resist instability through excess capacity. It survived by prediction, feedback, and timely correction.

Later tragedies would reveal the complementary lesson.

In the Challenger and Columbia disasters, the mathematics of flight remained intact. What failed was not physics, but regulation at the organisational level. Signals were present. Warnings were issued. But deviation was normalised, feedback was discounted, and decisions arrived too late to matter. Control failed not because equations were wrong, but because constraint was ignored.

Apollo succeeded because it listened to its mathematics.

Life operates the same way.

D. The Modern World: Civilisation as a regulated system

Several defining technologies of the modern world, artificial intelligence, electric transport, and magnetic levitation, share a common mathematical architecture. They operate close to instability and persist only through continuous correction.

Modern large-scale artificial intelligence systems do not encode knowledge explicitly. They define objective functions and reduce error under constraint. Learning is not understanding; it is convergence. Performance emerges not from precision, but from regulation acting iteratively under noise.

Electric vehicles became viable not when batteries became abundant, but when battery management systems could estimate internal state, regulate charge and temperature, and keep operation within narrow stability bounds. Energy is not stored as surplus; it is managed as risk.

Magnetic levitation systems make this dependence explicit. Levitation is inherently unstable. Suspension exists only because deviation is detected and corrected faster than it grows. When feedback slows, the system does not degrade gracefully. It falls.

These technologies reveal a broader transition. Earlier infrastructures relied on margin: excess capacity, mechanical tolerance, and robustness through redundancy. Modern systems rely on feedback: prediction, continuous correction, and operation near critical thresholds. They are efficient, adaptive, and fragile.

When regulation holds, performance is extraordinary. When regulation fails, collapse is sudden.

Civilisation increasingly resembles a living system.

Concluding remark

These episodes do not extend the argument of this book. They test it under conditions where intuition fails and error is unforgiving.

In each case, mathematics does not guarantee control. It clarifies what control would require. It distinguishes delay from stability, correction from appearance, and resilience from mere endurance. It shows when action can still shape outcome, and when time has already narrowed the space in which correction is possible.

In life, as in history, structure alone is never enough. What matters is whether structure can be sustained, quickly enough, precisely enough, and long enough, to matter at all.

Glossary of Mathematical Terms

(as used in this book)

Amplification

A process by which deviation increases rather than being corrected. In systems governed by positive feedback, small differences grow over time. Amplification is essential for transitions and development, but destructive when unconstrained.

Bifurcation

A qualitative change in system behaviour caused by a continuous change in parameters. At a bifurcation, a system shifts regime: stability may give way to oscillation, chaos, or runaway growth. Cancer is treated in this book as a biological bifurcation.

Boundedness

The condition under which a system's variables remain within limits that preserve coherence. Health is defined mathematically as bounded behaviour, not optimisation or stillness.

Constraint

The set of structural rules that determine what a system can and cannot do. Constraint does not restrict meaning; it makes coherence possible. Without constraint, behaviour dissolves into randomness.

Control

The capacity of a system to preserve invariants under disturbance. Control depends on feedback, timing, and information. It is distinct from delay: control restores structure, while delay merely postpones deviation.

Delay

Temporary suppression of deviation without restoration of underlying structure. Delay can mimic control in the short term, but allows drift to continue. Much of modern medicine excels at delay without fully restoring control.

Diffusion

The mathematical description of how variance increases over time under random fluctuation. Even when average behaviour remains stable, extremes become more likely. Diffusion explains why ageing manifests as increased variability rather than uniform decline.

Drift

The gradual accumulation of small errors due to imperfect regulation. Drift does not imply failure of control, but loss of precision. Ageing is described in this book as drift under continued regulation.

Error

Deviation from expected behaviour. In living systems, error is inevitable and informative. Regulation manages error; it does not eliminate it.

Feedback

A process by which system output influences future behaviour.

- Negative feedback counteracts deviation and stabilises systems.
- Positive feedback amplifies deviation and accelerates change.
  The balance between these determines stability or instability.

Invariance

A property that remains unchanged under transformation. Invariance defines structure. Life persists not by preserving material components, but by preserving organisation across change.

Instability

A regime in which small perturbations grow rather than decay. Instability arises when feedback fails to constrain behaviour. Cancer is treated here as instability without control.

Margin

The distance between typical system behaviour and critical thresholds. Margin allows error without collapse. Ageing narrows margin; health preserves it.

Noise

Random fluctuation intrinsic to complex systems. Noise is not an external insult to life, but a condition under which regulation must operate. Noise enables adaptation, but accumulates consequence over time.

Optimisation

The process of maximising performance relative to a defined objective. Optimisation is often brittle. Living systems prioritise boundedness and resilience over optimisation.

Phase Transition

A sudden change in system behaviour caused by crossing a critical threshold. Phase transitions explain why systems governed by continuous rules can change abruptly. Cancer represents such a transition in biological systems.

Prediction

The generation of expected future states based on internal models. Regulation depends on prediction. Living systems act before confirmation arrives; correction follows error.

Regime

A region of system behaviour characterised by stable patterns of response. Systems may shift between regimes without changing underlying rules. Health, ageing, and cancer correspond to different regimes of the same mathematics.

Regulation

The set of processes through which a system detects deviation and counteracts it. Regulation is active, energy-consuming, and probabilistic. Life exists because regulation persists under noise and time.

Robustness

The ability of a system to maintain function under disturbance. Robustness often relies on redundancy and slack. Excessive robustness without regulation leads to rigidity.

Stability

The tendency of a system to return to bounded behaviour after perturbation. Stability does not require stillness; it requires effective feedback.

Threshold

A boundary beyond which system behaviour changes qualitatively. Approaching a threshold is not equivalent to crossing it. Many failures appear sudden because thresholds are crossed after long accumulation.

Time

The dimension in which error accumulates. Time does not cause failure directly; it allows noise and imperfection to compound. Ageing is the signature of time acting on regulation.

Variance

The spread of possible system states around an average. Ageing increases variance even when mean performance remains unchanged. Failure often occurs in the tails, not at the centre.

A Note on References and Evidence

This book is not a catalogue of results, nor a comprehensive review of any single field. It is a work of synthesis. Its aim is to trace a common mathematical logic across domains that are usually studied in isolation, rather than to reproduce the full technical literature of each discipline.

For this reason, references are used selectively. They point to foundational results, historically decisive moments, and well-established frameworks that anchor the argument, rather than to every contemporary development or competing interpretation. Where claims rest on widely accepted scientific or mathematical principles, citations are intentionally sparse. Where the argument moves into interpretation, synthesis, or extension across fields, this is made explicit in the text.

Clinical examples are drawn from professional experience and are presented for conceptual clarity rather than statistical generalisation. Historical episodes are referenced to primary sources or authoritative secondary accounts, not as exhaustive histories, but as demonstrations of how structure, regulation, and time shape real outcomes under constraint.

This book does not report new experimental findings, nor does it replace detailed technical treatments in mathematics, biology, or physics. Its purpose is explanatory rather than predictive. Equations are used to clarify structure, not to calculate outcomes.

Where the book advances an interpretive position, it does so openly. Where it relies on established knowledge, that reliance is indicated. The boundary between evidence and interpretation is not blurred, but acknowledged.

Mathematics disciplines argument by revealing what can persist, what must drift, and what will eventually fail. The references provided here serve the same function: to ground the discussion without overwhelming it, and to support clarity rather than accumulation.

To recognise these equations is not to escape them, but to understand when intervention still matters, and when it no longer can.

Notes

Chapter 1. Mathematics as Pattern

1. Emmy Noether, "Invariante Variationsprobleme," *Nachrichten von der Gesellschaft der Wissenschaften zu Göttingen, Mathematisch-Physikalische Klasse* (1918): 235–257.
2. Hermann Weyl, *Symmetry* (Princeton, NJ: Princeton University Press, 1952).
3. G. H. Hardy, *Ramanujan: Twelve Lectures on Subjects Suggested by His Life and Work* (Cambridge: Cambridge University Press, 1940).
4. G. W. Dunnington, *Gauss: Titan of Science* (Washington, DC: Mathematical Association of America, 2004).
5. John Stillwell, *Mathematics and Its History*, 3rd ed. (New York: Springer, 2010).
6. Steven H. Strogatz, *Nonlinear Dynamics and Chaos: With Applications to Physics, Biology, Chemistry, and Engineering* (Boulder, CO: Westview Press, 2014).
7. Katsuhiko Ogata, *Modern Control Engineering*, 5th ed. (Upper Saddle River, NJ: Prentice Hall, 2010).
8. W. Ross Ashby, *An Introduction to Cybernetics* (London: Chapman & Hall, 1956).
9. Hiroaki Kitano, "Systems Biology: A Brief Overview," *Science* 295, no. 5560 (2002): 1662–1664.
10. Grégoire Nicolis and Ilya Prigogine, *Self-Organization in Nonequilibrium Systems: From Dissipative Structures to Order through Fluctuations* (New York: Wiley, 1977).

11. Ernst H. Gombrich, *Art and Illusion: A Study in the Psychology of Pictorial Representation* (Princeton, NJ: Princeton University Press, 1960).
12. Hermann Weyl, *Symmetry* (Princeton, NJ: Princeton University Press, 1952).
13. Michael Baxandall, *Painting and Experience in Fifteenth-Century Italy* (Oxford: Oxford University Press, 1972).
14. John Shearman, *Only Connect… Art and the Spectator in the Italian Renaissance* (Princeton, NJ: Princeton University Press, 1992).
15. Leonardo da Vinci, *The Notebooks of Leonardo da Vinci*, ed. and trans. Edward MacCurdy (London: Jonathan Cape, 1938).
16. Martin Kemp, *Leonardo da Vinci: The Marvellous Works of Nature and Man* (Oxford: Oxford University Press, 2006).
17. Kenneth Clark, *The Nude: A Study in Ideal Form* (Princeton, NJ: Princeton University Press, 1956).
18. David Freedberg, *The Power of Images: Studies in the History and Theory of Response* (Chicago: University of Chicago Press, 1989).

Chapter 2. Music, Expectation, and Structure

1. David Forinash and Wolfgang Christian, "String Resonance," *LibreTexts*, accessed January 3, 2026.
2. Hermann von Helmholtz, *On the Sensations of Tone as a Physiological Basis for the Theory of Music*, trans. Alexander J. Ellis (London: Longmans, Green, and Co., 1885).
3. Reinier Plomp and Willem J. M. Levelt, "Tonal Consonance and Critical Bandwidth," *Journal of the Acoustical Society of America* 38, no. 4 (1965): 548–560.
4. "The Well-Tempered Clavier," *Encyclopaedia Britannica*, accessed January 5, 2026
5. Peter Williams, *Bach: The Goldberg Variations* (Cambridge: Cambridge University Press, 2001).
6. Peter Vuust et al., "Rhythmic Complexity and Predictive Coding: A Novel Approach to Modeling Rhythm and Meter Perception in Music," *Frontiers in Psychology* 5 (2014): Article 1111.
7. Simon L. Denham, "Predictive Coding in Auditory Perception: Challenges and Opportunities," *Philosophical Transactions of the Royal Society B* 375, no. 1791 (2020): 20190371.
8. A. Peter Brown, "Eighteenth-Century Traditions and Mozart's 'Jupiter' Symphony," *Journal of Musicology* 20, no. 4 (2003): 480–521.
9. Rajesh P. N. Rao and Dana H. Ballard, "Predictive Coding in the Visual Cortex: A Functional Interpretation of Some Extra-Classical Receptive-Field Effects," *Nature Neuroscience* 2, no. 1 (1999): 79–87.
10. Eugene M. Izhikevich, *Dynamical Systems in Neuroscience: The Geometry of Excitability and Bursting*(Cambridge, MA: MIT Press, 2007).
11. György Buzsáki, *Rhythms of the Brain* (Oxford: Oxford University Press, 2006).
12. Michael Breakspear, "Dynamic Models of Large-Scale Brain Activity," *Nature Neuroscience* 20, no. 3 (2017): 340–352.
13. Gustavo Deco and Morten L. Kringelbach, "Metastability and Coherence: Extending the Communication through Coherence Hypothesis Using a

Whole-Brain Computational Perspective," *Trends in Cognitive Sciences* 20, no. 9 (2016): 655–668.

14. Viktor K. Jirsa et al., "On the Nature of Seizure Dynamics," *Brain* 137, no. 8 (2014): 2210–2230.

15. Peter Dayan and L. F. Abbott, *Theoretical Neuroscience: Computational and Mathematical Modeling of Neural Systems* (Cambridge, MA: MIT Press, 2001).

Chapter 3. Regulation and Persistence

1. Walter B. Cannon, *The Wisdom of the Body* (New York: W. W. Norton, 1932).

2. Claude Bernard, *An Introduction to the Study of Experimental Medicine*, trans. Henry Copley Greene (New York: Henry Schuman, 1949).

3. Norbert Wiener, *Cybernetics: Or Control and Communication in the Animal and the Machine*, 2nd ed. (Cambridge, MA: MIT Press, 1961).

4. W. Ross Ashby, *An Introduction to Cybernetics* (London: Chapman & Hall, 1956).

5. Erwin Schrödinger, *What Is Life? The Physical Aspect of the Living Cell* (Cambridge: Cambridge University Press, 1944).

6. Ilya Prigogine and Isabelle Stengers, *Order Out of Chaos: Man's New Dialogue with Nature* (New York: Bantam Books, 1984).

7. Hiroaki Kitano, "Biological Robustness," *Nature Reviews Genetics* 5, no. 11 (2004): 826–837.

8. Uri Alon, *An Introduction to Systems Biology: Design Principles of Biological Circuits* (Boca Raton, FL: Chapman & Hall/CRC, 2006).

9. Naama Barkai and Stanislas Leibler, "Robustness in Simple Biochemical Networks," *Nature* 387 (1997): 913–917.

10. Alexander Raj and Alexander van Oudenaarden, "Nature, Nurture, or Chance: Stochastic Gene Expression and Its Consequences," *Cell* 135, no. 2 (2008): 216–226.

11. Herbert A. Simon, "The Architecture of Complexity," *Proceedings of the American Philosophical Society* 106, no. 6 (1962): 467–482.

12. Rajesh P. N. Rao and Dana H. Ballard, "Predictive Coding in the Visual Cortex: A Functional Interpretation of Some Extra-Classical Receptive-Field Effects," *Nature Neuroscience* 2, no. 1 (1999): 79–87.

13. Blaise Pascal, *Pensées*, trans. A. J. Krailsheimer (London: Penguin Classics, 1995).

14. Ian Hacking, *The Emergence of Probability: A Philosophical Study of Early Ideas about Probability, Induction and Statistical Inference* (Cambridge: Cambridge University Press, 1975).

15. Isaac Newton, *Philosophiæ Naturalis Principia Mathematica*, trans. I. Bernard Cohen and Anne Whitman (Berkeley: University of California Press, 1999).

16. D. T. Whiteside (ed.), *The Mathematical Papers of Isaac Newton*, Vols. I–VIII (Cambridge: Cambridge University Press, 1967–1981).

17. Richard S. Westfall, *Never at Rest: A Biography of Isaac Newton* (Cambridge: Cambridge University Press, 1980).

18. Georg Cantor, *Contributions to the Founding of the Theory of Transfinite Numbers*, trans. Philip E. B. Jourdain (New York: Dover, 1955).

19. Joseph W. Dauben, *Georg Cantor: His Mathematics and Philosophy of the Infinite* (Cambridge, MA: Harvard University Press, 1979).

20. José Ferreirós, *Labyrinth of Thought: A History of Set Theory and Its Role in Modern Mathematics* (Basel: Birkhäuser, 1999).

21. Tolkāppiyar, *Tolkāppiyam* (c. 3rd century BCE–3rd century CE), esp. *Eluttatikāram*.

22. George L. Hart, *The Poems of Ancient Tamil: Their Milieu and Their Sanskrit Counterparts* (Berkeley: University of California Press, 1975).

23. Kamil Zvelebil, *The Smile of Murugan: On Tamil Literature of South India* (Leiden: Brill, 1973)..

24. Joseph Needham, *Science and Civilisation in China*, Vol. 2: *History of Scientific Thought* (Cambridge: Cambridge University Press, 1956).

25. François Jullien, *The Propensity of Things: Toward a History of Efficacy in China* (New York: Zone Books, 1995).

26. Nathan Sivin, "Chinese Conceptions of Time," in *The Study of Time III*, ed. J. T. Fraser et al. (New York: Springer, 1978), 75–92.

27. Marcus Terentius Varro, *De Lingua Latina*, Book V.

28. Mary Beard, *SPQR: A History of Ancient Rome* (New York: Liveright, 2015).

29. Reviel Netz, *The Shaping of Deduction in Greek Mathematics* (Cambridge: Cambridge University Press, 1999).

Chapter 4. Mathematics and Ageing

1. Carlos López-Otín et al., "The Hallmarks of Aging," *Cell* 153, no. 6 (2013): 1194–1217.
   Introduces the canonical domains of ageing; this chapter reframes them as interacting regulatory processes rather than independent mechanisms.

2. Carlos López-Otín et al., "Hallmarks of Aging: An Expanding Universe," *Cell* 186, no. 2 (2023): 243–278.
   Updates the framework to emphasise interconnection, feedback, and systems-level coupling.

3. Thomas B. L. Kirkwood, "Evolution of Ageing," *Nature* 270 (1977): 301–304.
   Establishes ageing as a consequence of imperfect maintenance and trade-offs, rather than a genetic programme.

4. Leonard Hayflick, "Biological Aging Is No Longer an Unsolved Problem," *Annals of the New York Academy of Sciences* 1100 (2007): 1–13.
   Frames ageing as progressive loss of fidelity in maintenance and repair systems.

5. Mario F. Fraga et al., "Epigenetic Differences Arise during the Lifetime of Monozygotic Twins," *Proceedings of the National Academy of Sciences* 102, no. 30 (2005): 10604–10609.

6. Steve Horvath, "DNA Methylation Age of Human Tissues and Cell Types," *Genome Biology* 14, no. 10 (2013): R115.
   Demonstrates ageing as drift and dispersion in regulatory state rather than deterministic change.

7. William E. Balch et al., "Adapting Proteostasis for Disease Intervention," *Science* 319, no. 5865 (2008): 916–919.

8. Jasmijn Labbadia and Richard I. Morimoto, "The Biology of Proteostasis in Aging," *Annual Review of Biochemistry* 84 (2015): 435–464.

9. Ning Sun et al., "Measuring In Vivo Mitophagy," *Molecular Cell* 60, no. 4 (2016): 685–696.

10. Alexandra Bratic and Nils-Göran Larsson, "The Role of Mitochondria in Aging," *Journal of Clinical Investigation* 123, no. 3 (2013): 951–957.
Emphasises coupling, regulation, and variability over simple notions of mitochondrial failure.

11. Robert A. J. Signer and Sean J. Morrison, "Mechanisms That Regulate Stem Cell Aging," *Cell Stem Cell* 12, no. 2 (2013): 152–165.

12. Thomas A. Rando and Howard Y. Chang, "Aging, Rejuvenation, and Epigenetic Reprogramming," *Cell* 148, no. 1–2 (2012): 46–57.

13. Rumiana Bahar et al., "Increased Cell-to-Cell Variation in Gene Expression in Ageing," *Nature* 441 (2006): 1011–1014.

14. Ioannis Angelidis et al., "An Atlas of the Aging Lung Mapped by Single-Cell Transcriptomics," *Nature Medicine* 25, no. 10 (2019): 1696–1707.
Provides direct evidence that ageing manifests as widening variance and heterogeneity rather than uniform decline.

Chapter 5. Noise and Accumulation

1. N. G. van Kampen, *Stochastic Processes in Physics and Chemistry*, 3rd ed. (Amsterdam: North-Holland, 2007).
Canonical mathematical treatment of noise, diffusion, and accumulation in stochastic systems.

2. C. W. Gardiner, *Stochastic Methods: A Handbook for the Natural and Social Sciences*, 4th ed. (Berlin: Springer, 2009).
Provides formal grounding for variance growth under noise, including $\langle x^2 \rangle \propto t$ in diffusive processes.

3. Hannes Risken, *The Fokker–Planck Equation: Methods of Solution and Applications*, 2nd ed. (Berlin: Springer, 1989).
Establishes probability flow, diffusion, and accumulation in dynamical systems.

4. Michael B. Elowitz et al., "Stochastic Gene Expression in a Single Cell," *Science* 297, no. 5584 (2002): 1183–1186.
Demonstrates that biological noise is intrinsic rather than pathological.

5. Alexander Raj and Alexander van Oudenaarden, "Nature, Nurture, or Chance: Stochastic Gene Expression and Its Consequences," *Cell* 135, no. 2 (2008): 216–226.
Shows how stochasticity propagates and accumulates across biological processes.

6. Rumiana Bahar et al., "Increased Cell-to-Cell Variation in Gene Expression in Ageing," *Nature* 441 (2006): 1011–1014.
Direct evidence that ageing manifests as widening variance rather than uniform decline.

7. Carlos P. Martínez-Jiménez et al., "Aging Increases Cell-to-Cell Transcriptional Variability upon Immune Stimulation," *Science* 355, no. 6332 (2017): 1433–1436.
Confirms variance amplification and heterogeneity with age.

8. Herbert A. Simon, "The Architecture of Complexity," *Proceedings of the American Philosophical Society* 106, no. 6 (1962): 467–482.
Introduces hierarchy, near-decomposability, and timing constraints across scales.

9.   Hiroaki Kitano, "Biological Robustness," *Nature Reviews Genetics* 5, no. 11 (2004): 826–837.
     Explains how robustness and fragility coexist under accumulated perturbation.

10. Marten Scheffer et al., "Early-Warning Signals for Critical Transitions," *Nature* 461 (2009): 53–59.
     Shows why collapse appears abrupt after long periods of apparent stability.

11. Vasilis Dakos et al., "Methods for Detecting Early Warnings of Critical Transitions in Time Series Illustrated Using Simulated Ecological Data," *PLoS ONE* 7, no. 7 (2012): e41010.

12. Charles Darwin, *Journal of Researches into the Natural History and Geology of the Countries Visited during the Voyage of H.M.S. Beagle* (London: John Murray, 1839).

13. Charles Darwin, *On the Origin of Species by Means of Natural Selection* (London: John Murray, 1859).

14. Charles Lyell, *Principles of Geology*, Vols. I–III (London: John Murray, 1830–1833).

15. Janet Browne, *Charles Darwin: Voyaging* (London: Jonathan Cape, 1995).

16. Janet Browne, *Charles Darwin: The Power of Place* (London: Jonathan Cape, 2002).

17. Ronald A. Fisher, *The Genetical Theory of Natural Selection* (Oxford: Clarendon Press, 1930).

18. George R. Price, "Selection and Covariance," *Nature* 227 (1970): 520–521.

Chapter 6. Instability Without Control

1.   Douglas Hanahan and Robert A. Weinberg, "The Hallmarks of Cancer," *Cell* 100, no. 1 (2000): 57–70; and "Hallmarks of Cancer: The Next Generation," *Cell* 144, no. 5 (2011): 646–674.
     Authoritative foundation for cancer phenotypes; this chapter reframes the hallmarks as consequences of lost regulatory constraint rather than independent causal mechanisms.

2.   Sasi S. Senga and Richard P. Grose, "Hallmarks of Cancer - The New Testament," *Open Biology* 11, no. 1 (2021): 200358.
     Extends and refines the canonical hallmarks framework, incorporating de-differentiation, epigenetic instability, microenvironmental influence, and neuronal signalling.

3.   Mina J. Bissell and William C. Hines, "Why Don't We Get More Cancer? A Proposed Role of the Microenvironment," *Nature Medicine* 17, no. 3 (2011): 320–329.
     Demonstrates that tissue context and environmental constraint actively suppress malignancy.

4.   James E. Ferrell Jr., "Self-Perpetuating States in Signal Transduction: Positive Feedback, Bistability, and Memory," *Current Opinion in Cell Biology* 14, no. 2 (2002): 140–148.
     Biological grounding for feedback-driven amplification, bistability, and state persistence.

5.   John J. Tyson, Kathy C. Chen, and Béla Novak, "Sniffers, Buzzers, Toggles and Blinkers: Dynamics of Regulatory and Signaling Pathways in the Cell," *Current Opinion in Cell Biology* 15, no. 2 (2003): 221–231.
     Formal analysis of switching, bifurcation, and control architectures in cellular regulation.

6. Steven H. Strogatz, *Nonlinear Dynamics and Chaos: With Applications to Physics, Biology, Chemistry, and Engineering* (Boulder, CO: Westview Press, 2015). *Mathematical foundation for bifurcation, instability, and regime change, independent of biological specificity.*

7. Mel Greaves and Carlo C. Maley, "Clonal Evolution in Cancer," *Nature* 481 (2012): 306–313. *Shows prolonged stability followed by rapid diversification during cancer progression.*

8. Peter C. Nowell, "The Clonal Evolution of Tumor Cell Populations," *Science* 194, no. 4260 (1976): 23–28. *Foundational demonstration that tumour progression proceeds through discrete evolutionary transitions.*

9. Christoph Lengauer, Kenneth W. Kinzler, and Bert Vogelstein, "Genetic Instabilities in Human Cancers," *Nature* 396 (1998): 643–649. *Frames genomic instability as an accelerator of progression rather than a primary initiating cause.*

10. Robert A. Gatenby and Robert J. Gillies, "Why Do Cancers Have High Aerobic Glycolysis?" *Nature Reviews Cancer* 4, no. 11 (2004): 891–899. *Interprets metabolic reprogramming as adaptive behaviour under altered regulatory constraints.*

11. Vanessa Almendro, Andriy Marusyk, and Kornelia Polyak, "Cellular Heterogeneity and Molecular Evolution in Cancer," *Annual Review of Pathology* 8 (2013): 277–302. *Establishes heterogeneity and adaptability as emergent properties of unconstrained dynamics.*

Chapter 7. Phase Transitions in Living Systems

1. Steven H. Strogatz, *Nonlinear Dynamics and Chaos: With Applications to Physics, Biology, Chemistry, and Engineering* (Boulder, CO: Westview Press, 2015). *Provides the formal mathematical framework for bifurcation, instability, and regime change.*

2. René Thom, *Structural Stability and Morphogenesis* (Reading, MA: Addison-Wesley, 1975). *Foundational work on qualitative change under continuous rules and loss of structural stability.*

3. Per Bak, Chao Tang, and Kurt Wiesenfeld, "Self-Organized Criticality: An Explanation of 1/f Noise," *Physical Review Letters* 59, no. 4 (1987): 381–384. *Introduces systems that naturally evolve toward critical points without fine-tuning.*

4. Nigel Goldenfeld, *Lectures on Phase Transitions and the Renormalization Group* (Reading, MA: Addison-Wesley, 1992). *Establishes universality, scaling, and regime-independent behaviour near transitions.*

5. Marten Scheffer et al., "Early-Warning Signals for Critical Transitions," *Nature* 461 (2009): 53–59. *Demonstrates hysteresis, irreversibility, and delayed recovery after threshold crossing.*

6. Stuart A. Kauffman, *The Origins of Order: Self-Organization and Selection in Evolution* (New York: Oxford University Press, 1993). *Frames biological systems as attractor-based, regime-dependent dynamical systems.*

7. James E. Ferrell Jr. and Sang Hoon Ha, "Ultrasensitivity Part III: Cascades, Bistable Switches, and Oscillators," *Trends in Biochemical Sciences* 39, no. 12

(2014): 612–618.
Provides concrete biological realisations of bifurcation, bistability, and oscillatory regimes.

Chapter 8. Health as Dynamic Balance

1.   Walter B. Cannon, *The Wisdom of the Body* (New York: W. W. Norton, 1932).
Introduces homeostasis as bounded, self-correcting regulation rather than static equilibrium.

2.   Peter Sterling and Joseph Eyer, "Allostasis: A New Paradigm to Explain
Arousal Pathology," in *Handbook of Life Stress, Cognition and Health*, ed. Shirley
Fisher and James Reason (New York: Wiley, 1988), 629–649.
Reframes health as context-dependent regulation rather than deviation from a fixed set point.

3.   Hiroaki Kitano, "Biological Robustness," *Nature Reviews Genetics* 5, no. 11
(2004): 826–837.
Establishes margin, redundancy, and slack as fundamental properties of healthy systems.

4.   John M. Carlson and John Doyle, "Complexity and Robustness," *Proceedings of
the National Academy of Sciences* 99, suppl. 1 (2002): 2538–2545.
Demonstrates why highly optimised systems become fragile under perturbation.

5.   Marten Scheffer et al., "Anticipating Critical Transitions," *Science* 338, no.
6105 (2012): 344–348.
Shows that distance from tipping points is more relevant to health than absolute system state.

6.   Ilya Prigogine and Isabelle Stengers, *Order Out of Chaos: Man's New Dialogue with
Nature* (New York: Bantam Books, 1984).
Establishes irreversibility and time asymmetry as intrinsic to living systems.

7.   Roger Penrose, *The Emperor's New Mind* (Oxford: Oxford University Press,
1989).
Clarifies limits of computation, prediction, and control in physical and biological systems.

Chapter 9. Core Systems and Regulation

1.   W. Ross Ashby, *An Introduction to Cybernetics* (London: Chapman & Hall, 1956).
*Introduces the law of requisite variety and explains why regulation, rather than optimisation,
enables persistence.*

2.   Ilya Prigogine, *From Being to Becoming: Time and Complexity in the Physical
Sciences* (San Francisco: W. H. Freeman, 1980).
*Establishes irreversibility, far-from-equilibrium dynamics, and time as a generator of both
structure and decay.*

3.   Hermann Haken, *Synergetics: An Introduction* (Berlin: Springer, 1977).
Canonical treatment of order formation and dissolution under noise in complex systems.

4.   Leonard Hayflick, "The Limited In Vitro Lifetime of Human Diploid Cell
Strains," *Experimental Cell Research* 37, no. 3 (1965): 614–636.
Direct experimental evidence that persistence under regulation has intrinsic limits.

5.   Thomas B. L. Kirkwood, "Understanding the Odd Science of
Aging," *Cell* 120, no. 4 (2005): 437–447.
*Frames ageing as a consequence of trade-offs in maintenance and regulation rather than
programmed failure.*

6. Brian C. Kennedy et al., "Geroscience: Linking Aging to Chronic Disease," *Cell* 159, no. 4 (2014): 709–713.
Supports the view that ageing shifts the system's operating regime rather than producing isolated point failures.

Epilogue

Quantum Systems: Structure without Regulation

1. John von Neumann, *Mathematical Foundations of Quantum Mechanics*, trans. Robert T. Beyer (Princeton, NJ: Princeton University Press, 1955)
Demonstrates that quantum systems obey exact mathematical structure without feedback-based error correction or regulatory persistence.
2. Roger Penrose, *The Emperor's New Mind* (Oxford: Oxford University Press, 1989).
Distinguishes immutable physical law from biological persistence, memory, and self-maintenance.
3. Carlo Rovelli, *The Order of Time* (New York: Riverhead Books, 2018).
Provides an accessible account of time, irreversibility, and why quantum systems do not "age" in the biological sense.

Stephen Hawking: Structure Without the Body

4. Stephen Hawking, *My Brief History* (London: Bantam Press, 2013).

5. Kitty Ferguson, *Stephen Hawking: His Life and Work* (London: Bantam Press, 2011).

6. Charles Seife, *Hawking Hawking: The Making of a Scientific Celebrity* (New York: Basic Books, 2009).

Artificial Intelligence: Regulation without Life

7. Judea Pearl, *Causality: Models, Reasoning, and Inference*, 2nd ed. (Cambridge: Cambridge University Press, 2009).
Clarifies why optimisation, inference, and prediction do not constitute self-preserving regulation.
8. Stuart Russell, *Human Compatible: Artificial Intelligence and the Problem of Control* (New York: Viking, 2019).
Supports the claim that artificial regulation is instrumental, externally specified, and not identity-preserving.
9. Karl Friston, "The Free-Energy Principle: A Unified Brain Theory?" *Nature Reviews Neuroscience* 11, no. 2 (2010): 127–138.
Provides a formal bridge showing why prediction and regulation resemble life while still requiring biological embodiment and history.

Life as Regulated Persistence

10. Erwin Schrödinger, *What Is Life? The Physical Aspect of the Living Cell* (Cambridge: Cambridge University Press, 1944).
    Classic articulation of life as persistence against entropy through constraint and regulation.
11. Denis Noble, *Dance to the Tune of Life: Biological Relativity* (Cambridge: Cambridge University Press, 2016).
    Defends multilevel regulation and rejects gene-level reductionism in favour of systems coherence.

Appendix

Selected Sources for Historical Case Studies

Cryptography, War, and Regime Change

1. David Kahn, *Seizing the Enigma: The Race to Break the German U-Boat Codes, 1939–1943* (Boston: Houghton Mifflin, 1991).
   Authoritative account of naval Enigma, the 1942 blackout, and U-559; emphasises operational consequences rather than cryptographic detail.
2. Ralph Erskine and Michael Smith, eds., *Action This Day: Bletchley Park from the Breaking of the Enigma Code to the Birth of the Modern Computer* (London: Bantam Press, 2001).
   Primary-source-based history highlighting delay, timing, and information flow.
3. F. H. Hinsley et al., *British Intelligence in the Second World War*, vol. 2 (London: HMSO, 1981).
   Official history; particularly strong on convoy losses, information lag, and regime shifts.
4. Clay Blair, *Hitler's U-Boat War*, vol. 2 (New York: Random House, 1998).
   German perspective confirming that Allied routing effectiveness, not submarine disappearance, altered outcomes.
5. Alan Turing, "The Applications of Probability to Cryptography" (1941; declassified).
   Shows cryptanalysis as constraint reduction and probabilistic control rather than message reading.

Nuclear Physics and Threshold Phenomena

6. Richard Rhodes, *The Making of the Atomic Bomb* (New York: Simon & Schuster, 1986).
   Unmatched synthesis of physics, mathematics, and decision-making under uncertainty.
7. John von Neumann, "Theory of Detonation Waves" (1944), Los Alamos reports.
   Primary mathematical basis for implosion symmetry and shock convergence.
8. Robert Serber, *The Los Alamos Primer* (1943; declassified 1965).
   Explicitly frames nuclear implosion as threshold crossing under timing constraints.

9. Herbert L. Anderson, "Neutron Physics and the Chain Reaction," *Reviews of Modern Physics* 18, no. 3 (1946): 347–361.
    Establishes the branching-process nature of criticality.
10. Peter Galison, *Image and Logic: A Material Culture of Microphysics* (Chicago: University of Chicago Press, 1997).
    Shows how robustness and uncertainty displaced precision in wartime physics.

Guidance, Control, and Feedback Engineering

11. Margaret Hamilton, "Computer Got Loaded," NASA Apollo Experience Report (1971).
    Definitive account of why Apollo survived overload through mathematical prioritisation.
12. Don Eyles, *Sunburst and Luminary: An Apollo Memoir* (Boston: Fort Point Press, 2004).
    Detailed explanation of Apollo Guidance Computer software architecture and feedback design.
13. NASA, *Apollo Guidance Computer Technical Manual* (1966–1969).
    Primary documentation of state estimation, feedback loops, and corridor constraints.
14. Eugene Kranz, *Failure Is Not an Option* (New York: Simon & Schuster, 2000).
    Operational perspective confirming that mathematical control, not heroics, maintained stability.
15. Charles Stark Draper, "Inertial Guidance Systems," *Proceedings of the IEEE* 52, no. 12 (1964): 1725–1734.
    Establishes guidance as continuous correction rather than computation.

Artificial Intelligence

16. Norbert Wiener, *The Human Use of Human Beings* (Boston: Houghton Mifflin, 1950).
    Foundational distinction between regulation and meaning.

Energy Systems and Active Stabilisation

17. Jean-Marie Tarascon and Michel Armand, "Issues and Challenges Facing Rechargeable Lithium Batteries," *Nature* 414 (2001): 359–367.
    Establishes batteries as intrinsically unstable systems requiring active regulation.
18. Gregory L. Plett, *Battery Management Systems*, vol. 1 (Boston: Artech House, 2015).
    Confirms state estimation, bounded operation, and risk-management framing.

Magnetic Levitation

19. Hermann Haken, *Advanced Synergetics* (Berlin: Springer, 1983).
    Mathematical treatment of inherently unstable systems stabilised by feedback.

20. Eiichi Masada, "Maglev Technology and Linear Drives," *Proceedings of the IEEE* 91, no. 7 (2003): 1028–1040.
Confirms that levitation exists only through continuous control.

The equations do not change; only the regime in which they operate does.
Notation is introduced only where it clarifies structure. No symbol is required to be remembered.

Notation & Symbols

This book uses mathematical notation sparingly and conceptually. Equations are not presented as tools for calculation, but as anchors for structure. They identify relationships, limits, and regimes rather than yielding numerical predictions.

The symbols below are intended to clarify meaning, not to invite computation.

Core Symbols

$x(t)$

The state of a system at time t.
Used generically to represent behaviour, activity, or organisation evolving over time.

x_0

A reference or preferred state.
Often corresponds to regulated equilibrium, homeostasis, or functional set point.

t

Time.
Not merely a parameter, but a source of accumulation, delay, and irreversibility.

$C(t)$

Coherence at time t.
A qualitative measure of how well a system preserves functional organisation.

$R(S, t)$

Regulatory capacity acting on structure S over time.
Represents feedback, correction, and control processes that preserve invariants.

$N(t)$

Accumulated noise over time.
Includes stochastic fluctuation, uncertainty, error, and perturbation integrated through history.

Structural Concepts

Invariance
What remains unchanged under transformation.
The mathematical definition of structure.

Boundedness
The condition that system variables remain within permissible limits.
Used as the operational definition of health.

Feedback
Processes by which deviation influences future behaviour.
Negative feedback stabilises; positive feedback amplifies.

Delay
Temporal separation between deviation and correction.
Distinguished throughout the book from true control.

Dynamical Regimes

Drift
Gradual loss of precision under continued regulation.
Characteristic of ageing.

Accumulation
Integration of small errors over time.
Leads to widening variance without immediate instability.

Regime Change
Qualitative shift in system behaviour without change in underlying rules.

Phase Transition / Bifurcation
A threshold at which continuous parameter change produces discontinuous behaviour.

Interpretive Notes

- Equations are illustrative, not predictive.
- Variables represent relationships, not measurable clinical quantities.
- Mathematical forms are chosen for conceptual clarity, not biological completeness.

This notation is consistent across chapters and equations. No prior mathematical training is required.

The Equations of Life

www.ingramcontent.com/pod-product-compliance
Lightning Source LLC
Chambersburg PA
CBHW080048240326
41599CB00052B/17